Contents

Preface

Surprisingly many books on project management have been published in recent years. From being a rather unconsidered branch of construction engineering, project management techniques seem to have been adopted and promoted widely as a kind of panacea for all kinds of business ills. Which is more than a little curious, for (you have only to keep an eye on the press to see this) several important projects go wrong. The question is, why do they go wrong when there is so much information about on how to manage them so that they go right?

Most of the answer has, I think, little to do with project management techniques, but a lot to do with project policy making. Project managers, as such, seldom have much say in the decision making which leads up to projects being conceived and brought into existence. Many people get involved with, and influence, projects without having had much exposure to the *specialness* of managing projects. Many project managers get assigned to a project without having had much exposure to the ways in which the outside world can influence their project. The world being what it is, such uninformed influence is more often bad than it is good for the project. But few people – whether they are politicians, bankers, investors, businessmen, planners or, least of all, project managers whose heads are at risk – want a project to fail. Most people want – intend – their projects to succeed and (one hopes) would welcome some discussion of what influences project outcomes. This book is intended for these people, whatever their background. It is addressed, not to their qualifications, but to their motives.

This is reflected in the way the book is organized. The first few chapters deal with projects as part of *business* risk. The middle deals with project investment, and the management of risk through financial and contractual arrangements. The customary tools of the project manager's trade appear only near the end. So the book does not say much about project management techniques (that, in any case, is easy enough to find elsewhere). It is mostly concerned with mind-sets, attitudes to project management, i.e. the 'project culture' of the final chapter, because this, when appropriately chosen and diligently applied, is what makes for effective project leadership – leadership, that is, by politicians, investors, corporate business managers and others in making project policy decisions, as well as leadership by project managers in getting the job done.

'Project culture' is, I believe, applicable to rather wide areas of management. We can properly talk about *project-based* management whenever the job we have to manage is delimited by specific objectives to be achieved within a budget and a schedule. But the principle is one thing, the practice quite another. Most of my examples come from the process industries – oil, gas and petrochemicals in particular – and it is not always easy to adapt the culture of these industries to, say, repairing cars or converting a grocer's shop into a gents' outfitter. Big hi-tech industries can afford (you may say) the elaborate procedures and attendant bureaucracy that big hi-tech projects always seem to demand. Small businesses cannot. But the culture that seeks to plan, check, control, audit – at a *suitable* level of detail – is surely good for both. Your problem, and it is a real one, is how to choose the suitable level. Only you, the project leader, can do that!

I have used masculine pronouns throughout for the sake of simplicity, intending no discrimination against the many project leaders and managers who are women.

References are quoted in the text when they relate to a specific topic under discussion. More general references are listed in 'references and further reading'.

I am conscious of the old adage that it is a wise man who writes, but a fool who publishes. In deciding, nonetheless, to publish I have benefited immensely from the experience and insights of many colleagues, practitioners and students. I acknowledge their help with sincere thanks. Special thanks are due to those who contributed to the seminars on which much of the book is based: Dennis Burningham, Tony Dark, Derek Glenton, Dick Harris, John Hitchens, the late Tom Ingram, Ambrus Janko and Mike Putin. Thanks also to Justin Taylor, who drew the illustrations, to Rodica Sartorius for translations from the Russian, to my editor David Ross, to the Production Editor Wendy Rooke and to the copyeditor Adam Campbell, for their unfailing patience and good humour. Such errors, omissions and infelicities as remain are mine.

John Dingle
Oxford
September 1996

1

Business Strategy and Project Investment

1.1 Introduction: the aims of business

The leaders of a business organization typically express their intentions for the organization's future in a 'mission statement'. This sets out the long-term goals which they intend the organization to achieve. Mission statements are declarations of intent for public relations, such as 'We aim to be the best in our field', or 'Our goal is to contribute to the well-being of the people and the nation'. They tend to be long on affirmation, but short on meaning. Meaning is added by corporate planning, which works out the specific objectives which should be reached in a particular time frame, and defines strategies: broadly, what has to be done in order to reach the planned objectives. Somewhere along the line, intention and strategy coalesce into policies, which often include *projects* as important steps towards achieving the objectives and, hence, the mission of the business organization.

We can illustrate this process by comparing an extrapolation of historical business performance (i.e. a 'target' for future performance) with what is likely to happen if no new activity is undertaken. The difference between what it is hoped to achieve and the probable result of simply maintaining the status quo is the 'planning gap'.

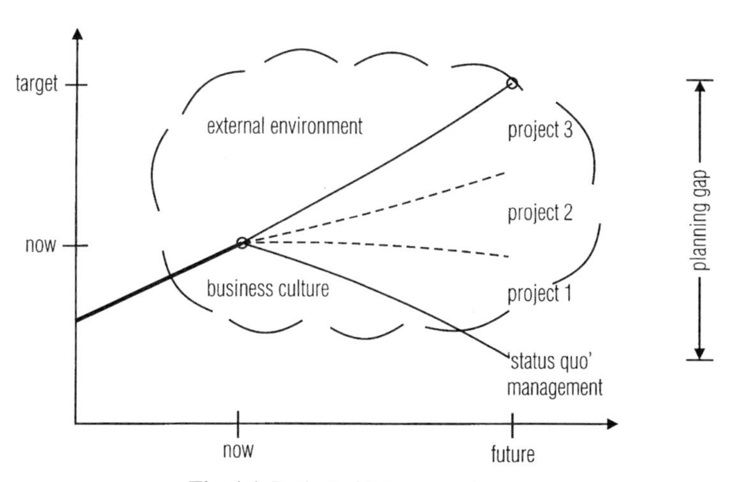

Fig. 1.1 Projects fill the planning gap.

The planning gap can be filled with one or more schemes (*see* Fig. 1.1). Each scheme has a definite start date, and a definite performance target to be achieved within a definite period of time for a definite amount of money. Each scheme is unique. Such a scheme is a 'project'. Projects create the assets on which the continued viability of the business depends.

Projects have features which distinguish them from other kinds of business activity and, in consequence, influence the ways in which project-related work may be directed and managed.

1.2 'Leadership' and 'management'

Leadership and management are not the same thing. Managers are, of course, expected to show leadership capabilitities, but they seldom direct the policies which shape the way in which projects are developed. Quite often, other people, whom I shall refer to as project 'leaders', exert considerable influence on the development of a project, especially in the early stages of planning. This can cause difficulties. We can see why by considering the differences between 'leadership' and 'management'.

Leadership, according to the Concise Oxford English Dictionary (COED), is to do with guiding, with directing actions and opinions by example, persuasion or argument, so as to cause others to go along with one to a certain destination. *Management*, on the other hand, is less glamorous. It is about getting things done (by fair means or foul, if we go along with the OED).

The characteristics attributed to 'leaders' are distinctly different from those attributed to managers. Leaders are said to be innovators, while managers are optimizers. Leaders seek change, which they see as improvement. Managers aim for continuity. Leaders exercise their influence through personality or charisma, while managers exercise theirs through authority. Leaders thrive in circumstances of ambiguity, uncertainty, rapid change and risk. They play things by ear. Managers strive for stability, establishing systems, standards and procedures. They play by the book. The one is a persuasive visionary, the other a technocrat.

We – supposing we are corporate power brokers – may be so convinced by these stereotypes that we forget the range of human capability and decide to appoint a project 'leader' to attend to those things which we believe are beyond our worthy project manager. Typically, those things relate to perceptions of the purpose of the project and its objectives. Characteristically, leaders adopt active, subjective or personal, attitudes to project objectives. Their ability to persuade influences expectations, and consequently changes perceptions of what is possible, desirable, practicable or essential. Characteristically, the project manager is likely to accept – or at least, acquiesce in – objectives or corporate aims the direction (and hence, implicitly, the practicality) of which has already been settled, taking the view that the time for fooling about with people's perceptions is past.

Because no project exists in isolation, its interfaces with the outside world are necessarily part of the project manager's concern. Interface management is notoriously difficult because of the value differences implicit in the very idea of an interface. The stereotypical manager will seek to reconcile these differences by compromise along some well-tried route. The stereotypical leader, on the other hand, will aim to manipulate along any route he can colour attractively. Not surprisingly, the leader's influence is likely to be most in evidence when business aims have to change, as they must, for example, under the influence of market forces.

1.3 Management tasks

Whether our destiny is to lead or to manage the achievement of corporate aims, we should appreciate the nature of the 'management task' which they imply. This may be, for instance:

- ensuring that plans and targets established by a higher authority are satisfactorily achieved;
- planning and administering so as to safeguard assets which we have inherited;
- ensuring the smooth running of complex operations;
- making profits so as to maximize the dividend payable to shareholders;
- optimizing operations in terms of customer loyalty to our own organization.

None of these tasks necessarily excludes any of the others, but their relative importance depends on the business organization's orientation to the market economy. Because the market economy changes, task priorities are liable to change. Consequently, project policies will change, and the portfolio of competences the project manager needs to do his job will also change.

Management tasks are carried out by people. We have already mentioned the project manager. This is the person responsible (and accountable) for managing the set of activities – the *scheme* – which we described in Section 1.1 above.

We may meet other project personalities, e.g. the 'project champion', who is appointed to safeguard and promote the project through its early development with a view to getting approval for the investment; and the 'project sponsor', who is the party for whom the project is undertaken. We must not forget the *shareholders*, who have equity interest in the project, and the *stakeholders*, who have other interests vested in project outcomes whether for success or failure.

The aspirations and expectations of these people do not always coincide and, indeed, may often be in conflict. If management is 'getting things done through people', and project management is 'getting things done *on time, within budget, and properly*, through people', then managing conflict will be an important aspect of the job.

1.4 Business change and the objectives of project investment

Handover to operations marks not only the completion of the project, but a step-change in the business. The objectives of investment in the project have been to bring this change about in the way that was planned, that is to say, in the way that the shareholders in the project expected when they authorized the investment.

These expectations arise from studies intended to define which project (among various options) and what timing will best fill the planning gap. But projects do not fill the planning gap neatly. In their early phases, before they are brought on-stream, they cost more than they earn, and it will be clear that the relationship between cost (project budget) and time (project schedule) will have a crucial effect on both shareholders' and stakeholders' perceptions of whether their expectations – of bringing about the desired business change – have been met or not (*see* Fig. 1.2).

The relationship between cost and time can be expressed roughly as an S-shaped curve, such as the right-hand curve in Fig. 1.3. Similar curves, which in effect summarize the project plan, become basic guidelines for the control of any project. The work carried out during the early stages of the project, which in the process industries typically amounts to some 10–20 per cent of the total project cost, determines most of the cost of the subsequent project work, i.e. 80–90 per cent of the total.

The greatest influence is exerted *before* the project is implemented, i.e. during the phase of studies. Later, the possibility of exerting strong influence falls off rapidly, as shown by the left-hand curve in Fig. 1.3. It is well-known that project changes made late in engineering usually result in dramatic increases in cost and/or delay in completing project work.

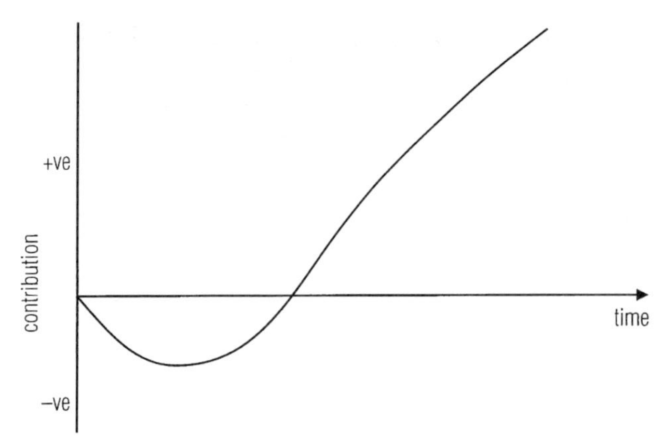

Fig. 1.2 The project's contribution to the business.

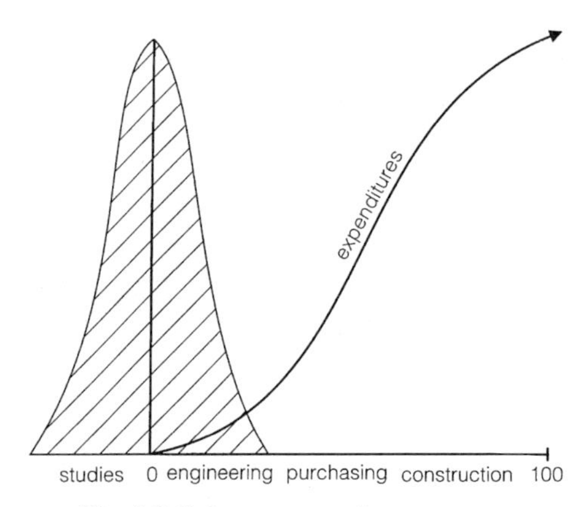

Fig. 1.3 Influence on project outcomes.

Studies cost little: from a few thousandths to a few per cent of total project cost. But it is the work done in studies which determines most of the cost of the subsequent engineering or design, which in turn determines most of the eventual cost of the project. It makes sense, therefore, to spend time, and money, on project studies, so as to reduce the uncertainties in the project before we spend time and money on implementing it. But people are often inclined to feel that the importance of work is measured by how much it costs. And they like to see something definite happen promptly, so they will often apply their influence to getting projects underway even though little effort has been spent on preparation. This is an example of where appropriate leadership (as defined above) is necessary.

1.5 Project characteristics

The COED defines a project as 'a plan; scheme; a planned undertaking'. The Association of Project Managers (UK) says a project is 'a set of inter-related tasks that are undertaken by an organisation to meet defined objectives, that has an agreed start and finish time, is constrained by cost, and has specified performance requirements and resources'.

In the Project Management Institute's *Body of Knowledge* (USA), a project is defined as 'any undertaking with an established starting point and defined objectives the achievement of which clearly signifies the completion of the project'. And the British Standard *Guide to Project Management*, in final draft at the time of writing, defines a project as 'a unique set of co-ordinated activities, with definite starting and finishing points, undertaken by an individual or organisation to meet specific objectives within defined schedule, cost and performance parameters'.

Definitions from industry have their own flavour:

> A project is the provision, within a given time frame, of hardware item(s) required to achieve a set of specific objectives (Shell: *Engineering Management Guide*).

> A project is a specific task to be completed to a specification within an agreed time and an agreed budget (BP: *Introduction to Project Management*).

In this context, the specification of 'performance' includes specification of social (e.g. health and safety) and environmental acceptability.

From all this, the characteristics of a 'project' can be summarized as a plane triangle (in which the angles represent *cost, time* and *performance* or 'duty specification'), circumscribed by regulatory constraints (*see* Fig. 1.4). This simple model demonstrates another important feature: as with any (Euclidian) triangle, we cannot change one angle without affecting one or both of the others. So we cannot change, say, 'time' without affecting either 'cost' or 'performance' or both. Further, we should remember that the model has to be drawn against a background. The background – the 'colour' of any project – is *risk*.

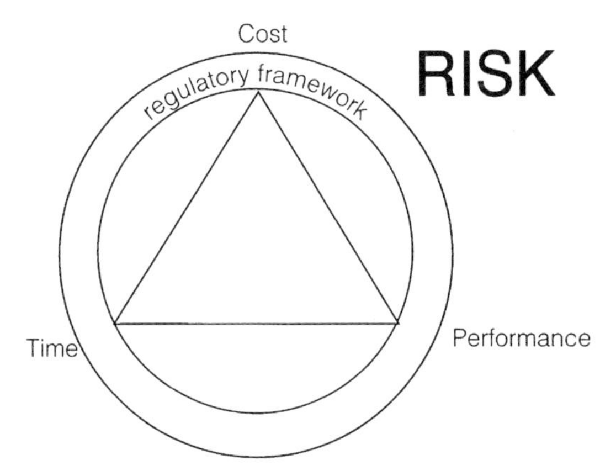

Fig. 1.4 Project characteristics.

In many industries, large projects are characterized by long execution time, technical complexity, major impact from influences outside the project, the once-only nature of the project which limits the application of past experience, high cost, high risk, and the need for an extensive project organization with necessarily complicated communication and information channels and procedures. A large industrial project requires the provision of engineering, construction, fabrication and operational services, together with the supply of goods and other services, particularly financial, required to complete it. Every project is different and will typically involve one or more technologies, requiring the participation of a number of companies.

The job of managing a project has been described since early times in terms of organizational skills and a capacity for wide-ranging forethought (Hori 1987). The project manager has to ensure that his project is completed within budget and on time. He has to lay plans, anticipate problems and see that objectives are met even when plans have to be changed. If the project is regarded as an interconnected system of activities with predetermined objectives, schedule and cost, project management is the discipline of conducting the necessary resources for execution of the project within the specifications of schedule, performance (e.g. quality, duty) and cost.

It follows that *project failure* means failure to meet project objectives as measured by the basic yardsticks of cost, schedule and performance specifications. Other criteria may also be taken into account, but these three remain the most accessible and the most fundamental. Remember, however, that the objectives and the yardsticks are those established at the time when project *implementation* is authorized, not those established at acceptance or some later date which may be more convenient or fortunate. Project failure, therefore, means failure to match the expectations of investors in the project, where 'investors' include all the parties having vested interests in the project. The fact that these expectations are the outcome of studies – paper replicas of the project – immediately ties the person responsible for project implementation into the preceding phases of the project cycle.

1.6 The project cycle

A convenient specification of the project cycle is that of the World Bank:

Identification – Preparation – Appraisal – Negotiation
Implementation
Operation
Post-project evaluation

To this we should add a final phase, 'close down', which includes decommissioning and the disposal of any remaining assets or liabilities.

There are several alternative definitions of the project cycle. They all describe a process of refining concepts until the risk in project decisions is acceptable, then realizing the project as an asset, and operating it for as long as it remains significant as an asset.

The first four of the World Bank's project phases are planning or 'development' phases:

- **Project identification** covers analysis of development strategies for the business (or, in national projects, the economy) as a whole, and identification of projects which seem to fit into and support these strategies, and which pass *prima facie* tests of feasibility.
- **Project preparation** covers examination of alternative means of achieving the notional objectives of projects identified in the preceding stage, and carrying out more detailed studies, a crucial feature of which is the preparation of cost estimates of increasing accuracy and confidence.
- **Project appraisal** means the comparison of the alternatives defined in the preparation stage, taking into account technical, institutional, economic and financial factors. This implies that all alternatives can be put on a comparable basis. Appraisal includes choosing the preferred project (or projects). This implies one or more decision rules for making the choice and investigating how sensitive the choice is to a variety of possible changes which may affect the project, singly or in combination. Such changes constitute risks that the preferred project may turn out to be less attractive than expected. It follows that we also need to consider what steps should be taken to eliminate or reduce these risks.
- **Negotiation** covers the organization and putting in place of relationships – usually contractual, legally enforceable, relationships – between the parties who together will supply the resources needed to ensure the success of the project, i.e. that it will meet its objectives

within its constraints so as to match the shareholders' (and, ideally, the stakeholders') expectations, as determined in the preceding stage.

After all this, *implementation* (building the project and putting it to work) and *operation* (keeping it working) are commendably straightforward, and *post-project evaluation* is simply to take a cool look at the project once the dust has settled and it has been working long enough to have established 'normal' operation, in order to see what was done well and would be done again in similar circumstances, and what could have been done better. Post-project evaluation is neither a post-mortem inquest on a dead body, nor a witch hunt. It is an opportunity to learn from successes as well as from mistakes.

The *close down* phase – a relatively recent addition to project cycle definitions – represents the belated recognition that a project has a finite useful life, and that it should not be simply abandoned when it ceases to be an asset, but should be removed and any residual value or liability liquidated.

1.7 Project success or failure: 'preconditions' and checklists

It is the work of the first four phases which mainly concerns us here, for it is in planning the project's development that the project leader's influence has most potential for good or bad. However, the planning of this 'planned undertaking' cannot be reduced to an algorithm such that a 'correct' solution is guaranteed when the 'correct' data are input.

How then to proceed? When we cannot rely on rules of thought, we often fall back on rules of thumb. They are derived from past experiences, from case histories. We appreciate that these are particular, not general, and that they may be distorted by hindsight. They are nonetheless the beginnings of a discipline for activities which are difficult precisely because they lack the discipline of structure.

A very detailed study of case histories of large projects lists a number of 'preconditions' for success (Morris & Hough 1986). They are in effect attitudes or dispositions of the management mind which favour a successful outcome. If they are present, it does not mean that the project is bound to succeed, any more than their absence means the project is bound to fail. In the latter case, however, failure is more likely.

Some of these preconditions are as follows:

- comprehensive and clearly communicated project definition;
- effects of external factors on the measures of project success clearly recognized;
- potentially harmful effects of urgency properly appreciated;
- good design/technology management, especially where there is technical uncertainty or complexity;
- good planning, clear schedules and adequate back-up strategies;
- full financial analysis of all project risks undertaken;
- support obtained from the relevant authorities;
- community factors properly considered and controlled;
- good, positive client, parent company and senior management attitudes, interrelationships and commitment.

This is by no means an exhaustive list, nor is it in any particular order, but it is instructive to compare it with the characteristics of our 'leader' and 'manager' stereotypes, against the background of the characteristic features of projects which we discussed earlier. We see the need for the attributes of both 'leader' and 'manager', and we see that the preconditions constitute a kind of scene-setting for the project manager in front of which he must play his role. Whether he appears as the hero or the villain of the piece depends very much on the set.

If preconditions help to get us in the right frame of mind, checklists help us to see what we have to do. Very many checklists of project planning activities exist. One of the most useful is published by the United Nations Industrial Development Organisation (UNIDO 1991). It is arranged as a workbook, so that completion of each section in succession builds up a definitive project appraisal report.

In order to be useful, a checklist must be complete: it must not omit any item that should be considered. Having in mind the range of project types with which we are concerned, we should not be surprised if a checklist, even if restricted to a particular project type – say, building a pyramid, or a petrochemical plant – grows to an excessive length when we try to make it exhaustive. And even if we are satisfied with the completeness of our checklist, we will find we have still not solved the other main problem which exercises managers, namely, how to prioritize action on each of the checklist items when the available resources are less than infinite.

1.8 Project pitfalls

This is perhaps an appropriate place to summarize the main areas of work demanding leadership and management in the light of the project development process as discussed so far. They are:

- clarifying project objectives;
- defining project work scope;
- developing strategies for design, financing, contracting, implementation, operation and eventual close down;
- setting realistic targets for cost, time and performance;
- putting in place organization, information and communication systems to mobilize and facilitate control of necessary resources, taking note of the primary resource – people.

Consider these project objectives:

- to generate wealth by adding value to raw materials;
- to raise standards of living by creating employment;
- to create opportunities for raising the skill level of the workforce;
- to improve health and welfare;
- to raise national consciousness and improve social cohesion.

Some or all of these objectives may be more or less achieved by, for example,

- a steel mill
- a chemical factory
- an urban centre
- a pyramid.

The extent to which each project satisfies its sponsors' ambitions – economic, social or political – is unlikely to be the subject of unanimous, stable or even objectively informed judgement. Nor should we suppose, if we are project managers, that wrangling of this sort is not our concern.

Earlier, I defined project failure in terms of expectations, as measured against criteria such as budget, schedule and performance specifications. Expectations are sometimes unrealistic. If they are, they have unfortunate consequences for the project and the project manager. For example, a contributory cause of unrealistic expectations is unwillingness to spend enough time – because this time is expensive – on making estimates the accuracy of which is suited to the decisions that have to made *at that stage in the development of the project.* Often, this results in decisions being made on the basis of preliminary, over-optimistic, estimates.

Furthermore, for capital-intensive projects, where capital charges are a major component of project cash flows, the problem of optimistic estimates is made worse by a project selection decision rule which chooses *maximum*, rather than *robust*, returns. In fact, any project selection which neglects the robustness of the choice under conditions of risk or uncertainty is vulnerable to the optimism of project planners – most notoriously, projects based on supposed economies of scale.

Since, in any project, people are the most important resource, we should at least touch on a few pitfalls in the field of 'organization', for this is where 'people problems' abound. Three examples are discussed below.

ORGANIZATIONAL STRUCTURES

Businesses of all kinds commonly structure their organizations in ways such that people report to superiors in a functional hierarchy. In principle, everybody knows who his boss is, and has the same boss for all aspects of the job (e.g. doing it, getting paid for doing it, etc.). The hierarchy supports the concept of business stability and continuity, and everybody should feel happy and secure.

Large projects are often carried out by a 'task force'. The task force lasts only as long as the project. Its boss is the project manager, who recruits people to the task force as and when he needs them. They are then responsible to the project manager for their performance within the task force, though they usually remain the administrative responsibility of their boss in the corporate organization.

Obviously, conflicts will arise between the business hierarchy and the task force. They must be resolved without damaging either of the organizational structures, or (too much) the egos of the people involved. But note that some social cultures cannot easily come to terms with the notion, implied by the task force structure, that people may have dual lines of responsibility. Also, some people get more worried by insecurity than they get excited by the challenges of work in a project task force, so reintegration into the business organization may be a problem at the end of the project.

PROJECT PROCEDURES

Procedures are needed to guide compliance with normal ways of doing work, i.e. the ways of doing work upon which everyone's project expectations are normally based. Procedures also assist timely and effective communication, and the recording of decisions and actions. The importance of communication in all phases of the project has been stressed before and will be discussed in detail in Chapter 7. It is sufficient to say here that the procedural aspects of communication inevitably generate immense amounts of documentation (which is also a good reason for trying to reduce the amount of paper, without reducing its utility), and if documentation management proves to be inadequate, then project procedures, no matter how excellent in themselves, will also fail.

Quite apart from this, project procedures will be strained if project objectives are changed, e.g. by market-induced changes or a decision to 'fast-track' the project (i.e. to accelerate the project schedule). A consequence of fast-tracking is that *normal* ways of doing work are disrupted. Project decisions and the work that follows from them stop being mainly sequential and have to be run largely in parallel (e.g. fabrication may begin before design is complete). The *normal* process of getting vendors to make competitive bids may have to be abandoned in favour of purchasing from preselected suppliers. Corporate management may have to accept that the project manager's authority must extend beyond its *normal* limits. These consequences should be, but often are not, considered *before* the fast-tracking decision is made.

PROJECT MANAGER'S AUTHORITY

This brings us to a central issue, namely the project manager's authority to carry out his responsibilities. The authority conferred on him is what makes him *accountable* for the performance of his job. He is responsible for mobilizing sufficient, suitable and timely resources, and because these resources are primarily people, he is responsible for motivating them and for ensuring that they, as well as he himself, are fully committed to the project objectives. In short, he is responsible for providing leadership within the scope of the project, and is accountable for the success of the project.

Examples of pitfalls such as these occur in most people's experience of business life. They help us to see how both leadership and management are necessary to achieve business aims. In the project environment, with its severe constraints and tightly specified criteria, leadership and management are not only necessary, but must also work in harmony. How this may be achieved, with the help of the 'project culture', is the subject matter of the following chapters.

1.9 Summary

To summarize: businesses must face the challenge of change; business leaders can respond proactively to the future by initiating projects to fill the planning gap which they foresee. Projects represent the fulfilment of shareholders' and stakeholders' expectations (i.e. the orderly bringing to reality of the concept of filling the planning gap). Investing in projects is consequently a strategy for reducing the uncertainty (or risk) in meeting the challenge of business change.

But when the business environment is subject to rapid change, project requirements – which derive from shareholders' and stakeholders' expectations – may change drastically. If they arise during project implementation, the consequences for the success of project can be serious.

The success of a project, from the point of view of the shareholders and stakeholders, will be assessed in terms of expectations at *completion*, not expectations at authorization, when the project's objectives were established. This fact is what makes it important for project leaders and managers to appreciate each other's role in responding proactively to changes in the business environment.

A project-based approach to business strategy uses project management methods and project-related ways of thinking – the so-called 'project culture'. Adopting the project culture can help us to improve the way in which we manage many different kinds of business activity, namely, all those kinds which can be characterized as 'projects'.

2
Project Development

2.1 Getting started

This chapter discusses the development of projects in the business environment. It is concerned mainly with planning and strategy issues which arise in the early stages, i.e. the time when studies are carried out with the aim of refining concepts and preparing for the realization of the business objectives embodied in the project.

This is the time when, because information is usually vague, contradictory or non-existent, many dubious decisions may be taken. It is a time when we would do well to reflect, in preparation for decision making, on the environment in which we will have to 'get things done on time, within budget, and properly, through people'.

CORPORATE PLANNING AND PROJECT STRATEGY

Corporate planning has two main aspects:

1. planning for the overall future direction of the business;
2. developing new business ideas from an initial concept through to commercial operation.

The second aspect is the one which most affects projects.

Typically, a project comes into being through a series of activities such as the following:

- a commercial opportunity is identified
- a general concept of a project is developed
- feasibility studies confirm the concept's commercial viability
- the project is defined
- design and construction
- commercial operation.

Real life is not as tidy as this. Various factors which change with time affect commercial viability, e.g. product values, raw material costs, actions by competitors, technological developments. Projects can be put on hold, sent back for re-definition, or dropped altogether.

However, once into the design and construction stage, the project is a real (rather than a paper) exercise. Few projects are stopped when this stage has been reached. But completing a project does not mean that its commercial viability is guaranteed, or that an adequate return on the investment will be achieved. Some projects are effectively abandoned without ever seeing significant commercial operation, and some limp on supported by the crutch of subsidies.

In addition to such dramatic failure, there is plenty of evidence that a significant proportion of projects go wrong in the sense that they fail to come up to the expectations on which investment was authorized. They are delayed, overrun their budgets, or simply do not work according to their specifications.

Why do projects fail? In the previous chapter, we saw that most of the total project costs will be decided by what is done in the early stages of development, and that these decisions are taken against what we might call a 'backcloth' of uncertain factors capable of influencing project outcomes. We categorized these uncertainties into (1) the external environment, and (2) the business culture (refer back to Fig. 1.1).

The *external environment* comprises all the exogenous factors which bear on the project but over which project people have little or no influence, and therefore no real possibility of control. Examples are changes in legislation, currency exchange rates, the activities of competitors and market consumption patterns. Of course, it may be possible to anticipate these factors and take timely and appropriate action to mitigate their effects on the project – sometimes even to benefit from them – but this falls far short of exercising control.

The term *business culture* refers to all the attitudes, mind-sets, customary practices and opinions within the business organization which colour (or condition, although they do not actually determine) corporate decision making. Project people may be able to influence these quite extensively, though they are rarely able to control them fully. Examples are attitudes to risk-taking; the corporate view of its role in, say, making profits; generating employment or responding to social or environmental pressures; opinions about the function of new technology in business diversification, and so on.

The external environment and the business culture, constituting the backcloth for project development, affect our judgements and therefore our decisions about developing the project.

One very important – often crucial – element of the project backcloth is taxation policy. The merit of the project as an investment opportunity depends very much on how it fits into the organization's tax position. Thus, as well as contributing to tax liability, the project may attract grants or subsidies if its location and timing are aptly chosen. Taxation is a very complex issue, whether we consider it falling on the organization as a whole, or on an individual project. Projects often go ahead or are killed off as a direct consequence of the impact of taxation policy. Because of its complexity and importance, taxation specialists should be involved at an early stage in project development. Appendix 1 provides an outline of the implications of fiscal measures for project development.

PLANNING FOR FUTURE BUSINESS: FORECASTING

It is difficult to imagine any kind of plan which is not based on a forecast.* Since time immemorial, people have sought a model of the future in order to benefit from fitting into it, or so that they can take steps to change with advantage to themselves or their kin. Indeed, the main reason for making a forecast may be to put oneself in a position to falsify it.

While the underlying importance of forecasts for business, including project-planning, is clear enough, we should note one important reservation. It is the axiom, which we neglect at our peril, that there are no facts about the future – there are only assumptions. The more distant the future

* In this section, 'forecast' is used indiscriminately to include statements such as 'it will rain tomorrow', and 'if present trends continue unchanged, rain is likely tomorrow'. The latter is, strictly speaking, a 'projection' and is safer than a forecast, but if your concern for verbal safety is that strong, you are unlikely to succeed as a leader.

we consider, the more dubious our assumptions are likely to be, and the more likely we are to leave them implicit, when they are likely to be overlooked as our work progresses. We must take care that all assumptions in forecasts are exposed to the light, i.e. made explicit.

Forecasting methods

While there are very many forecasting methods, they fall into only two classes: *judgemental* and *quantitative*. Judgemental forecasting relies on the intuitions, opinions and the experience of individuals (or the consensus of groups of individuals), and seems to be more frequently used in practice despite the obvious drawbacks. Quantitative methods have their roots in statistical analysis, which confers a measure of respectability, and may account for the fact that while managers are usually more at home with judgemental forecasting, they are more satisfied with quantitative methods.

An important procedure in judgemental forecasting is the DELPHI technique. In essence, this consists of asking topic experts for their opinion on the questions you wish to resolve, collating their answers, rephrasing the questions if necessary, and then recycling the collated answers until a consensus is achieved. A number of procedural variations have been developed, all intended to reduce the subjectivity of the expert responses. A frequent problem is that the appropriate experts who are accessible are our own in-house experts (the others all work for the competition!), so their opinions on a new project are unlikely to be entirely disinterested.

Among the important types of quantitative forecasting are those based on time series of statistics and those based on modelling. Time series methods aim to identify significant relationships between important variables and time, so that a time-based extrapolation can be used to forecast future values for the variable(s). An example is the use of business cycles to forecast future levels of business activity.

Modelling methods attempt to synthesize past relationships in sets of equations between variables. If the model so constructed produces the outcome actually observed in the past, it may be used to generate future outcomes or to test the consequences of change in one or several input variables. An example is the UK Treasury model of the national economy.

Judgemental and quantitative methods may be combined. Issues affected by social, cultural, political or large-scale technical developments are often forecast in this way, and the forecasts applied to strategic planning. All-embracing forecasts of this kind are the basis of '*scenarios*', i.e. hypothetical situations constructed so as to encourage managers to prepare 'what if' strategies. To be effective, scenarios must be realistic and credible (to line managers, rather than to planners). More than most forecasts, they depend on the existence of extensive and well maintained data bases.

Forecasting for business

Business, some would say, is simply about buying cheap and selling dear. Whichever way we see it, business depends on having a good idea of future demand for our product, and the cost of producing it and selling it in a future market.

Demand forecasting is where the trend of previous demand and an understanding of the factors which determine demand form the bases of our forecast, and where we will have to use our judgement in interpreting the forecast. But where do the data come from? The main sources, in order of increasing sufficiency and suitability of the data for a particular project, are business publications, multi-client studies, in-house data bases and specially commissioned research. This list is also in order of increasing cost – now, as ever, accurate and credible information has to be paid for.

We need to satisfy ourselves as to the following:

- Are the data sufficient and suitable, accurate enough for the kind of forecast we intend to make?
- Is the method we have in mind appropriate, especially in relation to the time-scale of the forecast?
- Does the method allow us to account for lags in the relationships between variables?
- Does it oblige us to make assumptions (such as using dummy variables) which, while making for mathematical elegance, ultimately reduce the credibility of the forecast?
- When we choose a variable (e.g. per capita GDP) to explain changes in another (e.g. demand for energy or plastics), is the direction of cause and effect unambiguous?
- Are the independent variables which we choose so closely correlated themselves that we cannot identify their separate effects on the dependent variables?

Cost and price forecasting is one of the most hazardous areas of forecasting, not least because we sometimes assume that 'price' is the same as 'cost plus a profit margin', and therefore that prices and costs always follow similar trends.We can see the flaw in this assumption as soon as we remember the existence of our competitors, unless we are used to living in a cosy cartel.

The market tends to ignore both extrapolations and models. It establishes both costs and prices in response to perceived demand (one person's cost is, sooner or later, another person's price). This has consequences for the data available to us for forecasting:

- Price reporting systems – an important source for data – do not usually treat each product uniformly, because the structure of each product market is different. This means that the relevance and timeliness of price data vary from one product to another.
- Data tend to autocorrelate so that forecasts become self-confirming, while actual trade is random.
- Each deal in the market is to some extent special, so that it is difficult if not impossible to make a statistical model which will retain its reliability in a fluctuating market.
- Data are often quoted in ranges, such as 'high/low prices'. The actual price is seldom at the midpoint of the range, though the midpoint is often supposed to be its location.
- It is impossible to get unambiguous data on contract prices because of the many modifications to what are nominally standard contracts of sale.

We all know that, for many products, price and demand are intimately connected. But the *elasticity* of demand with respect to price is not well understood, so that it is difficult to forecast the profit likely to result from a particular change in an existing price/demand relationship. (A new project may itself cause a change in the price/demand relationship for its product.)

Usually a new project boosts demand, especially if improved productivity allows prices to be reduced while sales revenues increase. But in some market situations, a substantial price reduction may have little effect on boosting demand, while quite a small increase in price will reduce demand considerably. This probably explains why, in oligopolistic markets, attempts to increase market share by price cutting tend merely to bankrupt all the suppliers (*see* Fig. 2.1).

Finally, we have to account for non-price factors which may influence demand. Demand for some products responds to the market's perceptions of product performance (which may be quite inexplicable) and to factors such as brand loyalty. While this is rather obvious for 'performance' products, it is also true to some extent of commodity products. Typically, the result is a non-linear relationship between price and demand, which considerably confuses the forecaster's best endeavours (*see* Fig. 2.2).

There is a great deal of literature on forecasting for business planning. Major industries, such as the petroleum industry, have been notable practitioners of the art, though perhaps with a little

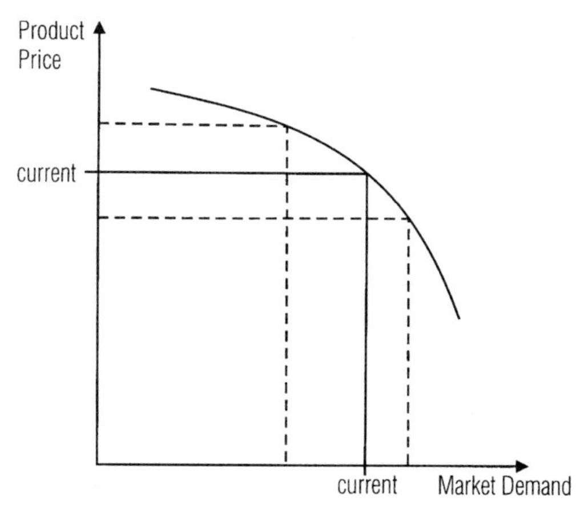

Fig 2.1 An anomalous price–demand relationship. Sales revenue is represented by the area of the rectangle under the curve. Neither increasing nor decreasing the product price is likely to increase revenue. Speciality products sometimes show this anomaly.

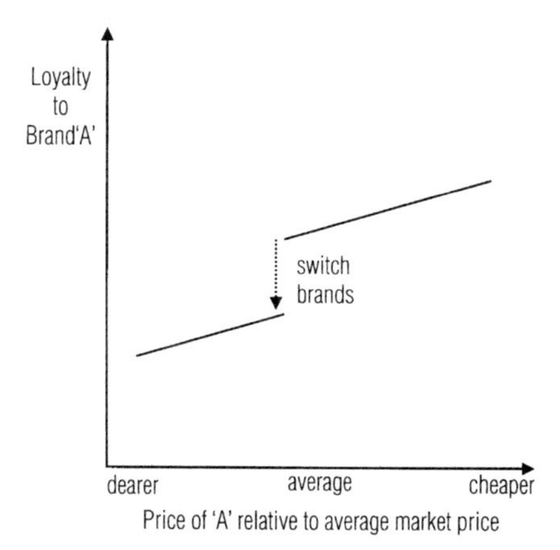

Fig 2.2 A market price/market loyalty relationship. Customers remain loyal to their preferred brand only up to a point. They may then switch to a cheaper brand. 'Performance' products sometimes show this anomaly.

less enthusiasm now for esoteric formal techniques than some years ago, when the world seemed a more stable place. This perhaps emphasizes that industry managers, while noting the forecasts put out for them, must rely on their own judgement when the key decisions about the future have to be made.

2.2 Project planning

INTRODUCTION

'Project planning', as used here, refers to:

- identifying a potential commercial opportunity;
- developing the general project concept (preliminary studies);
- confirming commercial viability (feasibility studies).

In principle, there is no limit to the number of investment projects that can be devised. But there is no point in acquiring rooms full of studies if the organization does not have the resources or willpower to implement serious projects. Many projects studied at this level will be discarded and will never proceed further.

Screening 'possible' projects against clearly specified criteria for acceptability produces a short-list of 'preferred' projects for more detailed study. Early studies should nonetheless be undertaken with an eye on the underlying technical, financial and contractual strategies.

The *technical* strategy is the set of plans (and back-up plans) detailing how the physical project deliverables are to be designed, constructed and put into operation in conformity with definite schedules, budgets and specifications. The *financial* strategy is the set of plans (and back-up plans) detailing how to mobilize financial resources in the appropriate amounts, currencies, and at the appropriate times to ensure the planned realization of the project deliverables. The *contractual* strategy is the set of plans (and back-up plans) detailing how to ensure that all the resources demanded by the project are properly brought together and coordinated to achieve the project's objectives.

These strategies must be integrated (itself another strategy!). As the project is developed, the various sets of plans will be more completely defined, but even at the earliest development stage, when the plans are vague, the strategies must not be neglected.

The more detailed aspects of project development will be dealt with later. Here, we will consider some of the aspects of early development, where the seeds of problems can be sown ready to grow into future project failure.

THE MARKET AND THE COMPETITIVE ENVIRONMENT

The market, in one way or another, is the most important influence on project development. A free market, according to economists, is supposed to provide an efficient mechanism for balancing through prices, supply and demand. This presupposes that all the players in the market have equal and simultaneous access to all significant information about the market, and that no player cheats.

There are several prerequisite conditions for markets to function in the way supposed by economists:

- efficient communications, so that the market players can get at and use market information;
- efficient banking systems, so that they can optimize payment for their transactions;
- comprehensive and unambiguous regulatory systems, so that players understand clearly what they can, and cannot, transact;
- transparent, unbiased monitoring systems to enforce the regulations and prevent abuse of the market.

To the extent that these prerequisite conditions are absent, the market becomes 'a place set apart for men to deceive and get the better of one another' (Anacharsis, *circa* 600 BC).

However, in a freely competitive market there will be some winners and some losers, so there is a built-in tendency for the market players to become unequal. Then the basic prerequisite conditions do not operate equally for all the players. There will be a tendency for the winners to pick the best options (because they will have acquired the means of getting better information, better communications, they can avoid regulations, etc.). Moreover, in the freely competitive market, winners need to demonstrate that they can continue to win. There will be a tendency to favour quick profit-taking at the expense of quality improvement (which takes time and commitment to build), so 'efficiency' comes to mean book-keeper's efficiency, and industrial efficiency tends to diminish. Attempting to move from centralized planning to a market-oriented economy and privatizing state-run industries both provide opportunities to observe the influence these tendencies have on the business environment.

Nonetheless, markets stimulate competition. Competitive forces of one kind or another influence all new investment projects. Hence, the wise project planner will try to estimate the strengths and weaknesses of the competitors, and identify the specific merits that the proposed project is supposed to have, for example, in technology, in raw material costs or in logistic advantages, and will take care to avoid overoptimism.

As competition becomes more highly developed, markets tend to become more risky, and investors naturally seek projects which show higher rates of return to compensate for the increased risk. This has an important consequence for project development. Outside the state sector – where we suppose for the time being that profit is not an issue – private and corporate investors will tend to favour short-term projects or those which increase the productivity of already existing facilities (so-called 'asset sweating' projects). Projects which employ innovative technology will be out of favour, and there will be pressure to 'fast-track' design and construction, and to eliminate refinements which do not contribute directly to earning power.

Recently, enthusiasm for development based on the market economy has been moderated by the feeling – perhaps not yet a belief – that free competition may lead to ecologically and socially unacceptable outcomes, and that development should therefore be constrained to what is sustainable in the long term. This feeling also carries implications for the project backcloth, for example, that:

- the market is not well suited to deal with long-term problems;
- in a consumer society, the market is driven by demand (=*I want*); in a sustainable development, the market should be driven by *need*;
- in a market economy, the application of technology aims to maximize throughput; in a sustainable development the application of technology should aim to maximize efficiency.

On the other hand, poor countries can claim that they need economic growth at a high (non-sustainable) rate in order to achieve conditions where sustainability becomes socially and politically acceptable. In other words, different levels of economic development demand different solutions in project development.

Sentiments like these are beginning to appear in corporate mission statements, and are likely to exert increasing influence on project development in future, as they become woven into the backcloth for the generation and screening of project concepts.

RESOURCE, LOCATION AND MARKET SURVEYS

A 'resource survey' studies the availability and suitability of resources for the project, particularly raw materials, utilities, services and people. While the need for the right quantity and quality of material resources is obvious, these features in respect of human resources are often neglected. Examples where neglect can cause future problems are in labour – or specialized skill-intensive manufacturing projects; and in large-scale construction projects in remote areas, or areas where other projects have depleted the normal labour pool.

Most commercial projects are based on the utilization of local resources and/or the supply of accessible markets. In any case, planning needs good information on resources and markets, and it is a good idea to begin these studies very early.

The choice of project location affects both the resources available and the potential markets, and is obviously a key factor in developing the project. Location studies focus on statutory requirements, logistics (of inputs to, intermediate products within, and outputs from, the project facilities), site conditions including infrastructural requirements, and factors to do with the project's impact on the local community, such as employment, safety and the disposal of effluents.

Thorough surveys on complex issues cannot be undertaken quickly or cheaply. Nonetheless, published information may be available which can at least provide some background for preliminary evaluations, though it is unlikely to be sufficiently specific or detailed to be useful for more than that.

The first requirement for a market study is that it should contain good information on the market as it is at the present time, and on how that position has been reached. Next, it should identify specific future developments as far as practical, e.g. planned new production capacity and future consumer patterns. Demand forecasts should allow for the production economics of key suppliers, including the cost of transporting and distributing products to the market.

In planning a survey, it is useful to draw up a checklist of the items to be covered. The checklist becomes a part of the study specification and will help with planning the development of the project.

PROJECT LOCATION

Choosing the location of the project site is closely related to market issues, notably in the case of processing or manufacturing projects, or projects involving major logistic activities. A basic question is whether to locate the project close to raw materials or close to product markets. The choice will affect marketing policy, and hence the project's expected earnings. Marketing policy is linked to pricing policy (especially in relation to 'transfer' pricing and the influence that the project location may exert on non-price factors – such as customer loyalty – within pricing policy) and to policy relating to product inventories held at intermediate points in a product 'chain' (because of the 'pipeline' effect where relatively small demand fluctuations affecting the upstream end of the chain can become relatively large fluctuations further downstream).

Social and political factors also influence the choice of project location. Among these are the views – supportive or otherwise – of Government, local authorities and the neighbouring community about the project as an economic and industrial development: issues to do with the project's infrastructural needs and impact (e.g. transportation, health and welfare, housing, schooling), and ecological and environmental considerations.

Choice of location will also be influenced by, for example, availability of raw materials, utilities and services, people (skills and work culture as well as numbers) and related facilities for training in operations and management, and regulations affecting design, construction and operation.

2.3 Generation and screening of project ideas

Any organization which is alert to its future always has a long list of possible projects. The list can always be made longer by imaginative 'brainstorming'. We want more ideas on the list than we have resources available for implementation, because some will be discarded in the screening process. So how do we choose which projects to investigate further?

The principle is to establish a ranking of the projects, based on definite criteria, e.g.:

- availability of raw materials
- extent and timing of market demand
- complexity or suitability of technology

- scale of investment required
- availability of suitable skills
- synergy with existing operations
- availability of the information needed to make firm decisions.

These criteria can be used as indicators of project viability before detailed calculations and special studies are made. Ranking is carried out by setting up a table with ideas on one axis and criteria on the other. Taking one criterion at a time, each project concept may be awarded points. When all the criteria have been considered, the points are added up and a preliminary ranking established. Take care to choose criteria so that the concepts they represent do not overlap. If, in the list above, 'synergy with existing operations' actually means familiarity with existing technology, then technology is going to be counted twice if 'suitability of technology' is also a criterion. This double counting will bias your selection procedure.

The method of scoring also needs to be thought through. The criteria you choose are unlikely all to have the same weight. Some will be more important than others in the view of your organization – or, at any rate, in the view of individuals whose opinions may be decisive in the organization. So suitable weighting factors may be assigned to the criteria, giving more weight to the more 'important' criteria. (You will have noticed already how easy it is to rig the results of project screening exercises. Beware!)

PROJECT ECONOMICS

Having identified a promising project idea, decisions have to be made on production capacity, location of facilities, phasing of implementation, and similar basic questions which affect the project economics. This set of decisions could be described as the 'project design' although this should not be confused with the engineering work, which is done much later. In fact it is important to avoid getting into too much technical detail at too early a stage, as this can easily obscure the fundamental decision making. On the other hand, neglect of key technical concepts sows the seeds of glitches (i.e. sudden, unexpected irregularities or malfunctions), which may later cause a good project to evolve into an expensive disaster.

Data on plant inputs, outputs and order-of-magnitude investment are essential when developing the project design. There is no point in trying to appraise a project without this information. Appraisal data are often provided by plant contractors and process licensors, who evidently have some degree of vested interest in the outcome of the appraisal. Consequently, data obtained from these sources should be treated with some caution.

Although a detailed economic appraisal belongs to the final phase of the feasibility study, the basic principles should be known and kept in mind from the earliest stages of project development. The price realized for the outputs of the project is one of the most critical variables in any appraisal. However, it is also usually the least controllable and least predictable element.

A method for examining price trends and industry profitability is the analysis of cash margins (i.e. the unit cost of producing an additional unit of product). The basic assumption is that the plant operator would be justified in continuing operation so long as his cash margin is positive. There will be a range of cash margins among the competing organizations due to different efficiencies and circumstances. If prices dip to the point where some competitors have negative cash margins, they may choose, or be forced, to stop operating. The consequent output reduction will tend to raise market prices, which may benefit the survivors. Some plants regularly shut down and reopen in response to such price signals. Others do not have this degree of flexibility, and thus are vulnerable to price volatility.

By now we will have assembled a great deal of information focused on project economics – the estimated costs and benefits of achieving objectives which themselves are not yet fully defined. We will be aware that this information is more or less incomplete and more or less unreliable. But we have used it, together with a liberal dose of judgement or prejudice, intuition and hope, to short-list some projects which we believe our organization might be prepared to accept. At this stage, a dangerous pitfall is 'championing'. None of our embryonic projects is yet strong enough for promotion by an ambitious individual eager to steal a march on the competition. His time will come, but after more detailed study.

SCOPE AND PURPOSE OF FEASIBILITY STUDIES

Once some preliminary appraisals of the project concept have been made, the next step is to undertake a feasibility study. Quite often, a series of studies will be required for a major project. The aim is to show (1) the conditions in which the project could achieve its objectives, and (2) that in achieving its objectives, it is likely to be economic (or commercially viable). Initial studies usually point to important questions which must be analysed and answered in subsequent more detailed studies. If an early study indicates a potentially worthwhile project, subsequent studies are undertaken to:

- review and update resource and market studies
- investigate the proposed location and potential alternatives
- review the available technology
- provide an outline design and cost study
- indicate infrastructural requirements
- prepare draft project implementation plans
- provide financial and economic analyses
- incorporate design, negotiating and operational philosophies.

Eventually these studies lead to a paper replica of the project – a model – which the organization considers credible. This model is the one which, we hope, will be approved for project definition.

2.4 Project definition

PROJECT DEFINITION STUDIES

Crucial decisions are made during project definition. They greatly influence the time and cost of physical completion of the project, and its subsequent viability. For example, deciding the choice of process technology determines about 80 per cent of the total costs of the project. Definition studies take a lot of time and effort, and are therefore expensive. This is why eager project promoters sometimes try to do without them or to reduce their scope, arguing that the feasibility studies should be enough. This argument can be resolved only in terms of the degree of uncertainty which the investors in the project are prepared to accept when they commit their money, and what they are willing to pay to bring uncertainty down to the acceptable level.

Four key areas for detailed study are:

- technology selection
- duty specification of the project facilities
- the project implementation plan
- financial and contracting plan.

These provide the foundations for drawing up the programme for acquiring the project plant. While they are discussed in later chapters, we should look here at 'bid strategy', i.e. the strategy for obtaining project resources from the various potential suppliers. For present purposes, we shall refer to all suppliers as 'contractors', including in this term others such as consultants, licensors and, in general, vendors of equipment, materials or services. (Note that, even when all necessary project resources are to be obtained in-house, contract-like arrangements need to be set up with the in-house departments which will function as 'contractors', and a bid strategy must be considered.) In the meantime, the question of who is to be invited to bid needs to be decided.

BID STRATEGY

For a major project, completely open bidding wastes time and may increase risk. Usually the contract will be for engineering, procurement and construction (EPC) or turnkey, and much effort is needed to evaluate the technical differences between bids. Completely open bidding encourages proposals from hungry suppliers with little relevant experience, in which keen pricing often merely reflects inability to perform.

This is why it has become common practice to prequalify suppliers as the basis for drawing up a short-list of those who will be invited to bid. This procedure is welcomed by reputable contractors, who will sometimes decline to prepare a full proposal unless bidding is restricted to a reasonable number of companies.

Prequalification, if properly conducted, will save time and effort at the final bid stage and does not involve contractors in expensive design and estimating work (for which the client organizations ultimately pay).

Prequalifying a short-list of potential suppliers is like project concept screening, and we can use the same method (see Section 2.3 above), applying suitable criteria and weights. Only the short-listed suppliers will receive invitations to bid (ITB). Our bid strategy should also look forward to the way we intend to evaluate the eventual bids.

Bid evaluation requires experience, foresight and attention to detail. Contractors are adept at leaving obscure yet significant omissions, which are designed to elicit 'extras' and 'variations' during the contract. It is worth emphasizing at an early stage the importance of making sure that all of the necessary works are included. There is no point in building a production plant which lacks the necessary utility, raw material supply and product dispatch facilities. Also, the timely provision of suitable infrastructural works is crucial to ensure good progress in construction.

The main items to be checked in evaluating bids are:

- identification of differences in the scope of supply
- completeness of the information provided
- willingness to meet special requirements
- suitability of the design philosophy
- validation of performance claims
- provisions for revising bids
- interviewing key project staff.

Note that it will usually be necessary to ask for revisions to put all the bids on a comparable basis or to cater for circumstances not foreseen in drawing up the ITB.

2.5 Project implementation

Project definition studies look forward to the implementation phase. Having mobilized the contractors (or our in-house resources), we now have to attend to many activities in parallel. The question of contractor liaison is in itself quite complex. Now we should remind ourselves of our original objective: to establish a viable commercial operation. This depends primarily on people and on the skills which these people bring to the project.

If the project is one involving only addition to the capacity of an existing company then it is a relatively straightforward task to ensure that sufficient trained staff are available in time for plant commissioning. However, if you are faced with building up a whole new organization, covering all functions including production, marketing, accounting etc., more complex problems arise. In this case, staff formation will be a major preoccupation.

Marketing is a key function. It is possible that the project may have been set up on the basis of only one major sales contract, as is often the case with big liquified natural gas (LNG) export plants. But any significant project will demand that we develop a marketing plan and set up the necessary organization.

As project implementation approaches completion, staffing will have to be built up ready for start-up and operation. Plant start-ups can go smoothly if they are well planned, but it is wise to assume that there will be some problems, if only because of unfamiliarity with equipment.

Things can also go badly wrong, and relations with contractors often hit a low, during commissioning. Delays can mount up due to errors in design or faults in construction; spare parts may not be available and operating procedures may need to be refined. While it is difficult to generalize on how best to handle plant commissioning, clear definition of responsibilities between client and contractor is fundamental, especially when the client's personnel are responsible for operating the plant during performance tests.

One problem to be faced is the fact that far more people are needed to start up a new plant than the normal operating complement. Even if employment gain was one of the project objectives, there really is no point in having excess staffing. Overstaffing encourages slack work and standards, which can be very difficult to change and will tend to perpetuate in future projects, adversely affecting their viability.

On this question of future projects, the planning strategy has come full circle. The project will have had difficult and tense moments. People will have gained valuable experience. Jobs will have been created, and if the planning was right, there will be a flow of revenue to repay loans and provide for new investment in new projects. Otherwise . . .

2.6 Summary

Projects are conceived and come into existence as means of achieving the objectives planned in corporate strategy. They are means of managing changes that are essential to business growth, or even to survival.

Project development begins with forecasts of the business environment, which is largely determined by market competition. The earliest stages of project development inevitably involve the generation and screening of many ideas to assess their relative feasibility. The purpose of feasibility studies is to refine project concepts to the point where definition is complete enough for investors to risk their resources on one preferred project.

Throughout project development, our aim is to ensure that the project is actually *manageable*. Conditions for project manageability include having realistic objectives, and having sufficient and suitable resources in place to achieve those objectives. Manageability implies spotting problems and pitfalls and resolving or avoiding them *before* they become significant.

3
Management Decision Making

3.1 The process of decision making

Decision making is a process. It involves a number of steps which lead to an outcome. For example, in order to *decide* an issue, we need to:

- define the situation in which the issue arises;
- analyse it: what are the causes, constraints, risks in the situation, and what do we aim to do?
- what options do we have?
- analyse the options;
- choose the preferred option;
- act on the preferred option.

In management we rarely have complete, reliable information about the issues we have to decide. We usually have to use our judgement – which includes some intuition, perhaps based on experience, perhaps just on 'gut feel' – as well as on whatever quantitative methods are available to us. We are not likely to make perfect decisions, though we may aspire to make decisions of the sort which Herbert Simon (1981) called 'satisficing', i.e. the best decisions we can make in the circumstances as we perceive them.

Most management decisions have to be implemented through other people, so our decision making process should, so far as is practicable, encourage acceptance by the other people who are affected by our decisions. This means thinking about our approach to decision making as well as the methods we use.

This chapter does not go deeply into decision theory. For that you must consult specialized texts on management science and rely on the experts for advice in more technical areas. But the eventual decision will be yours, not the experts'. Certainly you should use technical aids when these are appropriate (and when you're sure you know their limitations and your own), but mostly you will need to develop a 'feel' for the soundness of your decisions.

3.2 Decisions about projects

EVALUABILITY

When we develop a project, we are always faced with assessing the pros and cons of possible schemes. This is the business of project appraisal or *evaluation.* A fundamental question is: can the project be evaluated in a meaningful way? The point is that, if there are any doubts about a project's 'evaluability', you can be sure there will be doubts about its manageability.

It has been suggested (Rutman, 1976) that an 'evaluable' project should have:

1. a clearly articulated programme
2. clearly specified goals and/or effects
3. a rationale linking the programme to the goals and/or effects.

We must remember that the basic 'rationale' of project evaluation is very simple. It is to show the difference between expected costs and expected benefits due to the project's coming into existence. We say that if the difference (benefits *less* costs) is positive – or more positive than some alternative way of investing our money – then we should go ahead with the project.

However, consider these four classes of commercial projects:

1. **Making new or better products to meet a market demand.** There seems to be no great difficulty in evaluating projects like this.
2. **Cost-saving projects.** Again, there seem to be no problems in evaluating projects which aim to reduce costs, and therefore to improve profits relative to existing methods of production. Types 1 and 2 are both evaluable.
3. **Maintaining old business.** This class is very difficult to evaluate by demonstrating an increase in profitability, unless it can be shown that failure to replace an existing asset will lead to the loss of existing profitability. The danger is that the replacement will merely continue to produce things for which demand is declining.
4. **No-return projects.** These projects have outputs for which there is no market value (or none that can be reliably quantified). Welfare projects are typical examples. Another example is expenditure on satisfying government legislation on, say, safety or environmental protection. In the latter case, one may build the increased costs into the evaluation of the overall scheme of which the project is part. But if that is not possible, we have to conclude that type 4 is non-evaluable.

PROJECT DEFINITION

Evidently a study itself represents some level of project definition. We may have:

- pre-feasibility studies
- feasibility . . . viability studies
- detailed studies . . . pre-engineering . . . engineering studies,

which provide increasing levels of definition, aimed largely at getting better capital cost estimates.

Improving the quality of capital cost estimates can be quite expensive. At the 'sharp end' of conceptualization we will be lucky to be more accurate than ±30 per cent of the eventual cost. Spending more on a 'front end' engineering study ought to give us ±10 to ±20 per cent accuracy. But if we want better than ±10 per cent, we will probably have to spend a lot more on preliminary engineering.

Estimates made early in project development – e.g. for feasibility studies – cost relatively little, yet these are frequently the very estimates on which managements aim to save money. This is a mistaken economy, for these estimates usually determine the major expenditures later on. Project managers should be alert for:

- corporate management attempts to up-rate cheap estimates;
- top management reluctance to pay for accuracy;

- proposals to save investment by reducing operational flexibility; the cheapest plants are rarely flexible, yet actual operating circumstances often demand more flexibility than planners expect;
- the 'study' level of accuracy, e.g. ±30 per cent on plant costs, when no evaluation is likely to be very robust.

TECHNOLOGY SELECTION

Closely related to issues of capital cost estimation in project definition are issues of technology selection. Technology evaluation at the early stages of study is a particularly difficult topic, for the data available may not provide a sufficient indication of qualitative – much less quantitative – aspects of producing on-specification products. The difficulties are increased if one considers technology acquisition within the framework of a licensing agreement or a joint venture. Technology development does not stand still. Technology evaluation tends to be a 'snapshot' view of the technology as it is now, rather than as it will be when the proposed plant is operational. So ways have to be found of ensuring the ongoing viability of the project and, of course, of the skills of the people who run it.

Consider a few basic issues:

- **Choosing technology.** This is usually done early in the study phase of project development. The choice of a technology actually commits about 80 per cent of operating costs and (what may be even more of a problem in some locations) a similar commitment to realistic spares and maintenance budgets.
- **Offshore.** Offshore process facilities are built upwards, not sideways as onshore. So changes in design, especially extensions in horizontal space, are not as easy to accommodate. The main issue is safety, but weight (which has to be lifted into place and properly distributed when in place), environmental and noise problems can be severe enough to affect safe operation.
- **Performance.** The information used to evaluate technologies often comes from the owners of the technology through licensors or contractors, and may get massaged on the way. Typically, performance data are given in three categories:
  - 'normal operation'
  - 'expected'
  - 'guaranteed'.

  We need to be sure which category of data is being used in our study. Also, note that the law of conservation of mass sometimes does not seem to apply to commercial process technology, where 'losses' may be understated or not mentioned at all!

  Detailed performance data cost money, sometimes in the form of 'looksee fees' from licensors, always for developing engineering data. The first stage of engineering information development is the 'process package', and it is worth observing that the extent and quality of information contained in what are sometimes called 'standard' process packages may vary considerably. Any shortfall of information for design will cost something to make up.

PROJECT IMPLEMENTATION

Planning the implementation of the project is, in effect, the keystone in the bridge between the project as a concept and the project as an operating production facility. The implementation plan involves making many important decisions.

Here, as elsewhere, techniques must be used intelligently if they are to be useful at all. In this, the techniques are no different from other tools used by craftsmen: not only the tools, but some experience of using them properly is needed.* The problem is that experience entails trial and error (it is a characteristic feature of engineering – and perhaps of decision making in general – that progress is only achieved *via* mistakes) and one does not want one's errors to be too many or too great. It follows that, although every project is unique, the experience of others making project decisions – and especially their mistakes – is extremely useful.

There are two common ways of getting at the experience of other people. One is to get them to write it down in the form of checklists of items that should be taken into account at each stage of project development and fully understood before going on to the next stage. Another way is to examine case histories of projects, and several of these are available. Some of them express conclusions also in the form of checklists, i.e. such as checklists of topics which are in effect the 'preconditions' of project success. The presence of such 'preconditions' does not guarantee success, though absence may more certainly dispose towards failure (Morris & Hough, 1986, see Ch. 1 references).

Checklists and case history conclusions look at projects from opposite points of view. Checklists imply a view from now to the future (the planned 'now' to the hoped-for project future). Case histories, by definition, provide a view of the past, with the benefit of hindsight.

It would be useful somehow to focus these perspectives together. If you do this by combing through the relevant literature, you will end up with a very long list of topics, plus a number of precepts which have to do with the way the checklist topics should be considered. Essentially, the precepts are about the appropriate attitude of mind to adopt at each stage of project planning. A constant theme in these precepts – i.e. in the interpretation of project definition checklists – is that of 'expectation'. This is not surprising, for everything we have to do in project development is to do with the future, and there are no facts about the future: only assumptions which colour our expectations.

An 'expectation' is the product of the value of a future outcome and the probability of its happening. If we have a lot of experience about a previously known situation, we are inclined to think that the probability of its continuing is high, though we may consider that this probability will decrease in the future. At a lower – but still reasonably high – assessment of the current probability, e.g. in an 'either/or' situation, we may expect the future probability to change quite suddenly, depending on the amount and reliability of the data we can collect, and on our judgement. But if the current probability is low, we may consider our future expectation to be more or less of a gamble, when our view of the 'utility' of the expectation becomes significant for the decisions we make. In any case, what we are trying to do is to choose the 'best' option among those which we perceive are open to us.

3.3 Choosing the best option: decision making tools

INTRODUCTION

We make decisions every day, and we believe implicitly that the decisions we make are good ones. Of course, sometimes outside events conspire to frustrate our intentions, and sometimes the information we have on which we base our decisions proves to be wrong. If we reflect on these

* Stafford Beer, writing about industrial operations, observed: 'There is no *rigorous* means of knowing which (of the many variables involved) matter. Indeed the importance of a particular variable in such a system is a question of degree, a question of judgement, a question of convention.' Beer was making the point that not only the numerical value of the variable changes, but its *structural relevance* within the system, changes with time. (Waddington, 1977).

facts, we are forced to the conclusion that decision making is perhaps more complicated than our everyday experience would suggest.

Large projects in a sense call for large decisions. We have to decide on the commitment of large resources: money, effort and time, and to be responsible for consequences of our decisions – consequences which may be quite far removed from the project itself. A natural gas project is a good example. It involves the exploitation of an important national resource. It will consume huge sums of money and manpower. It will affect the lives of many people, not only those engaged on the project, but many more who have nothing directly to do with it. It will affect the environment. Its impact will be felt for a generation or longer. In these circumstances, we shall naturally be very concerned about the quality and quantity of the information we need for decision making. While this is obvious in the case of decisions on a large scale, it is no less significant for all management decisions.

In addition to information, we find we use intuition in decision making. There is no doubt that 'intuitive rationality' forms part of the way in which decision makers operate. I mean by this – again following Herbert Simon – the part of arriving at a decision which is due to experience and specific expertise, yet which is hardly ever distinctly articulated. This is in contrast to the kind of rationality attributable to objective experiment and analysis.

In practice, and especially in connection with project decisions, we seldom have enough complete and reliable information to support wholly objective-rational decisions. The deficiency is made up from intuitive rationality, and we accept that the resultant decision will be less than objectively optimal (Albino, 1988).

Shortage of reliable information is most acute in the early stages of project development. It is the business of project preparation to remedy this shortage, and it is the business of project appraisal to put this information into a form in which it can be used to measure project expectations against some suitable yardstick and to compare them. It is only then that we can hope to make a decision which is amenable to constructive discussion and eventual acceptance as the basis for effective action. However, *information* in itself is not enough: we also need to have a *decision rule*.

A decision rule is the formal method we adopt for choosing one option among the various alternatives which have been examined, measured and compared. If, for example, we chose to test projects against some criterion such as net present value (NPV), and we had been careful to observe the rules for this kind of appraisal, we might make our decision rule: 'choose the project which shows the maximum positive NPV'. In the earliest stages of project development however, we are often painfully aware that we simply do not have enough firm data to support such a clear-cut decision rule.

INITIAL SCREENING DECISIONS

Suppose we are not yet ready to choose a project: we are merely trying to screen a large number of possibilities so as to produce a short-list of projects suitable for further investigation. Our objective is to begin the evolution of a strategy for development which will enable us to make the eventual appraisal cost-effectively.

In the previous chapter, we saw how we could easily score project alternatives against a number of criteria and apply weighting factors to arrive at a ranking order of preference. We also saw how easy it is to reach a spurious decision because of the dubious values we may put into the scoring and weighting factors.

A rather more refined method of decision making at this level, where we are obliged to rely heavily on qualitative information (and intuition), is the directional policy matrix (DPM) technique made popular by Shell some 20 years ago. It is a method of ranking projects in terms of strategic planning objectives by locating potential projects on a matrix, in which the columns represent opinion as to the future profitability of the relevant industry sector, while the rows represent opinion as to the project sponsor organization's competitive capability. The pattern of one's own, and the competitor's, projects in the matrix gives an indication of the preferred development strategy (Channon and Jalland, 1979).

DPM and similar approaches take us about as far as we can go without firm data. If we wish to be able to defend our project decisions in the usual corporate cut and thrust, we also need to consider other more quantitative methods of decision making. In what follows, only a few – though important – aspects of quantitative decision making are touched on, and these at a cursory level which, I hope, will nonetheless stimulate interest and help enhance the credibility of your decisions.

PAYOFF MATRICES

A payoff matrix is a device for displaying the options and the various factors which influence choice. For example, suppose you are on vacation in an unfamiliar part of the country, driving along and enjoying the view and the sunshine. Suddenly, your companion (I shall assume this is your wife) announces she is hungry. You pass a sign which says there is a transport cafe a few hundred metres ahead, but you know your wife dislikes this sort of establishment. You mentally construct a payoff matrix like the one in Table 3.1.

Table 3.1 'States-of-the-world' payoff matrix

Strategies	(a) There is another, delightful restaurant 5 km down a side road	(b) There is no other restaurant for miles
1. Stop at the cafe	Wife annoyed	Wife relieved, but sulks
2. Go on in hope	Everyone happy except your bank manager	Divorce threatened

In this matrix, I have called 'states-of-the-world' those outside events over which you have no control, but which you have to take into account. They are sometimes called 'external events', 'states of nature', or sometimes 'scenarios'. 'Strategies' are the actions which you, as decision maker, may choose. Each combination of a strategy with a state-of-the-world leads to a 'payoff', i.e. the result of using the selected strategy in the particular circumstances or 'state-of-the world'.

The matrix may be constructed with as many distinct states-of-the-world, and as many different strategies, as we can identify. But it is best to keep the matrix fairly simple, because multiplying states-of-the-world or strategies tends to produce overlaps, resulting in ambiguous payoffs and confusion.

Although the matrix displays all the possible payoffs for the given situation, it does not tell you which is the right choice to make. In fact, with few exceptions, there is no uniquely 'right' choice. (The main exception is where the same strategy can be applied to what are essentially the same circumstances, and where one can therefore choose the strategy which leads to the maximum expected long-term average payoff. This situation does not normally arise in connection with projects, where decisions are essentially 'one-off' decisions.) Consequently, we need to choose a decision rule to apply to the strategies available to us.

Some typical decision rules, with their conventional names, are as follows:

- *Maximin* – choose the strategy which shows the least costly 'worst' case, i.e. limit the liability of failure.
- *Maximax* – choose the strategy which shows the best payoff in the best case, i.e. maximize the benefits of success.
- *Minimax regret* – calculate the opportunity costs and benefits for each case and choose the strategy which involves the greatest loss of gain if foregone.
- *Best average payoff* – calculate the (unweighted) average payoff for each strategy and choose the strategy which shows the best average.
- *Best expected value* – calculate the probability-weighted average payoff for each strategy and choose the strategy which shows the best expected value.

These decision rules assume that we are using money as the term for quantifying each case. This assumption is not always valid (e.g. in the domestic example above) and even when used, it may have to be modified to take account of the 'utility' of money, a concept which we will discuss further in the section on 'utility' (p. 33). Nor is money necessarily the only term we should take into account. Sometimes other terms must be included in the matrix, which then becomes a 'multi-attribute' matrix.

However, even without detailed analysis, constructing a payoff matrix clarifies the problem by reminding us to attend to all the possibilities. Further, the matrix may immediately show up some strategies which are 'dominated' by others in the sense that there is no criterion by which the dominated strategies could be considered attractive. These can be discarded, thus considerably simplifying the problem.

DECISION TREES

Many project decisions are not 'once for all time', but form part of a chain of decisions in which earlier positions affect later ones. We often have to choose a strategy which will affect what we can do subsequently. If we get our earlier decisions wrong, it may be difficult, or impossible, to achieve our long-term objectives. This problem of sequential decision making is made more graspable by setting out a decision tree.

Suppose I have it in mind to look for natural gas in my back yard. I can choose to drill or not to drill. The latter costs nothing; the former costs some considerable sum, but may possibly lead to a discovery which will make me rich. If the discovery is a good one, it will make me very rich. Even if it is marginal, it will make me richer than I am now. On the other hand, I may find no gas, in which case I shall have lost the money I invested in drilling.

Suppose I work out the cost of drilling, and the revenues earned by finding gas. Suppose further that I get some geological data which allow me to establish the probability of the various possible finds. I can construct a payoff matrix as in Table 3.2 (in money units). If my decision rule is to maximize expected returns, I will decide to drill, because the total expectation from drilling (10) is greater than the expectation from not drilling. But if my decision rule is to minimize expected losses, I won't drill, because the possible loss of –60 is worse than not drilling at all.

However, I need to consider what I will do with the gas if I find it, i.e. what my subsequent decisions will be, and how they will affect my first decision (to drill or not to drill). These issues can be investigated with a decision tree.

The payoff matrix in Table 3.2 becomes the first part of the tree. Subsequent options to exploit the alternative outcomes of drilling then become branches of the tree, and tree branching can be continued to whatever degree of detail you think fit (*see* Fig. 3.1). As the number of decisions in

Table 3.2 Payoff matrix – drilling for gas in the back yard

	Cost	Revenue	Payoff	Probability[a]	Expected return
Don/t drill	0	0	0	–	0
or					
Drill: dry hole	100	0	0	0.6	–60
Drill: small find	100	200	100	0.3	30
Drill: big find	100	500	400	0.1	40
				1.0	**10**

[a] Note that the probabilities must sum to 1 to show that all possibilities have been accounted for.

sequence increases, the number of branches grows very rapidly. Even two sequential decisions with their respective possible outcomes can result in a number of branches which is extremely tedious to evaluate by hand calculation. There is, however, plenty of computer software available to ease this chore.

SENSITIVITY AND RISK

It is a simple, though also tedious, exercise to show that rather small changes in, say, market penetration would falsify the decision reached in the example discussed above. That is to say, the decision is sensitive to market revenues.

In analysing the *sensitivity* of a decision, we consider specific changes in input data, and recalculate the payoffs or the decision tree. Changes in market demand or the achievable price are frequently very sensitive areas. For large-scale projects which take a long time to construct, change in the effective discount rate due to inflation, exchange rate fluctuations and similar financial factors are also sensitive areas.

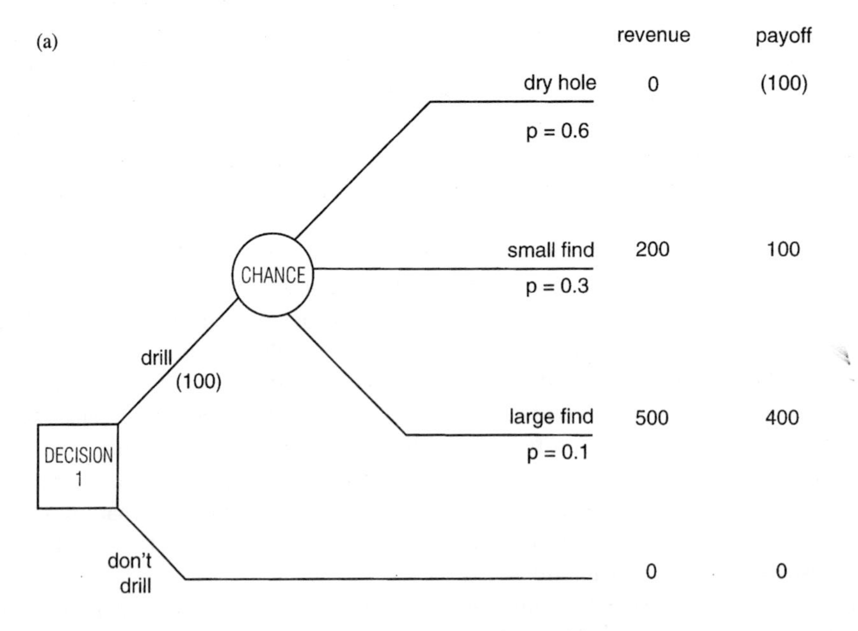

Fig. 3.1(a)

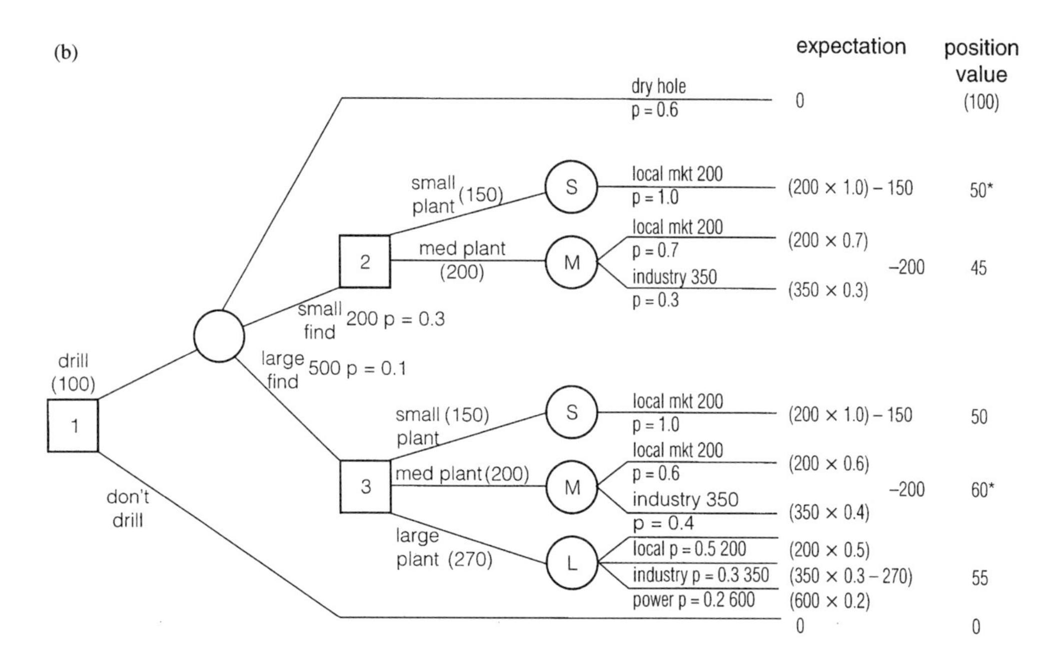

Expected Value of Decision 1 = 'Drill': (60 + 500)0.1 + (50 + 200)0.3 – 100 = 31

Fig. 3.1 Decision tree: (a) first phase, (b) second phase.

For a given type of project, it is usually fairly obvious which factors are likely to be most sensitive, and it is easy to run them through a computer (e.g. using a spreadsheet program) to estimate their influence on appraisal criteria. However, it is also easy to forget that disasters seldom come singly. The sensitivity of decisions should be tested for combinations of disadvantageous changes, as well as for each individual change. Clearly, you need to exercise judgement as to the likelihood of such combinations arising in practice.

In sensitivity analysis, we are in effect assuming the certainty of each case we consider. Alternatively we can assign a probability distribution to each case. This is a feature of 'risk analysis'. For example, instead of saying in an appraisal that the operating costs of our project will be 100$/mt of product, and testing the sensitivity of profits to an increase of say 10 per cent (i.e. to $110/mt), we can say that operating costs will be not less than 95 and not more than 115 $/mt (a uniform probability distribution); or, will be not less than 95$/mt, not more than 115$/mt and most likely $100/mt (a triangular distribution); or any other probability distribution for which we have evidence.

This works – apart from the explosive increase in computational power it requires – when (1) there is some evidence to support the probability values we assign to events (states-of-the-world), and (2) we actually know all the events which relate to the project. Most of the time, we are not in so happy a position. Either we know the events, but we don't know the probability of their occurrence (a situation called by analysts 'uncertainty'), or we know neither the events nor their probability (a situation called with wry humour '*partial* ignorance').

Conventional sensitivity analysis tends not to be helpful in these cases, because preferences appear to be very precariously balanced, and few of us have a very clear notion of the practical implications of series of probability factors. Sometimes it is possible to make the decision by reference to a decision rule which does not explicitly depend on probabilities, such as maximin, maximax or minimax regret. This may, however, involve *implicit* assumptions about probabilities, which may lead to a worse decision.

Using probability distributions in analysis produces outcomes which are also in the form of probability distributions. With these we can choose a statistical function of the distribution to represent the most likely outcome, or we can choose some other point in the distribution which represents a level of confidence which is acceptable (e.g. 'NPV will not be less than \$$x$ mn with n per cent confidence').

A sensible approach is to look for the least sensitive, i.e. the most *robust* choice: the one that is least affected by change. This should lead to a favourable payoff. The problem is then to define what is meant by 'favourable'. Usually we will look to use more than one decision rule, for instance:

1. NPV greater than x; and
2. capital exposure less than y.

Projects meeting both criteria would qualify as 'acceptable', while others would be considered 'unfavourable'.

Robustness is most useful when very many states-of-the-world have to be considered, and there is a high degree of uncertainty (or partial ignorance) associated with deciding the initial strategy. Where there are not many final payoffs, and the routes to them are fairly stable, it is often simpler to use the conventional decision tree approach, though a considered view of the robustness of an initial decision which is likely to constrain later options is always worth making.

Risk – the chance of an undesirable outcome occurring – becomes a greater problem when the information which is available to us about a risky situation is imperfect, i.e. is incorrect, incomplete or otherwise unsuitable. Being obliged to use *imperfect* information – because none other is available – is characteristic of management decision making.

The first step in managing risks is to identify them and to classify them by type. A simple, straightforward, initial approach is to classify risks as either 'physical' or 'financial' risks. We can then decide (in principle) for each type whether it is feasible to:

- *avoid* the risk (e.g. by adopting a different design, although in a project this may only substitute a different risk for the one we have avoided);
- *reduce* the risk (e.g. by design; by choosing appropriate hardware; by setting up suitable procedures; by special training);
- *transfer* the risk – or some of it – to another party (e.g. through contractual arrangements, or by insurance);
- *accept* the risk (e.g. by making contingency allowances, by in-company insurance,or by being prepared to pay for losses incurred).

The ranking of risks can be done in several ways, depending on the nature of the (imperfect) information available. In principle, risk ranking methods can use qualitative or quantitative information, or a mixture of both, and can incorporate statistical analysis (where adequate statistics are available) including 'fuzzy' data analysis. The main difficulties arise from:

- designing a suitable ranking scale which will adequately reflect possible views on acceptability;
- assigning weights to the various risk factors considered;
- the method of combining the weighted factors to arrive at the overall rating;
- choosing the criteria for overall acceptability.

However, the most important thing for the project decision maker to keep in mind is that it is not the risks which he foresees (and assesses) which cause trouble, but the risks which he does not foresee. We need, therefore, to cultivate risk *awareness* in our approach to project decision making.

UTILITY

Project decisions are about future events, so all the outcomes (payoffs) which we calculate will be expressed as 'expectations'. If we are considering money values, the expectation will be the product of the money sum involved times the probability of our receiving it. And we assume that the expectation of say $100 with a probability of 0.01 is the same as the expectation of $2 with a probability of 0.5.

This is very often not true, as can easily be shown by inviting people to participate in gambling games. Similarly, people's attitude to the risk implicit in future events is likely to differ, depending on whether the chance is one of gain or of loss. How can this feature of human judgement be taken into account in decision making? A method of doing so is found in 'utility' (or 'preference') theory.

Utility theory supposes that a continuous relationship can be found between a payoff and the decision maker's perception of its value (i.e. its 'utility') in the face of risk. A decision maker is said to be 'risk neutral' if this relationship is constant throughout the range of payoffs and perceptions (or 'utilities') considered. If the utilities tend to decrease with increasing money value, the relationship is one of aversion to risk: the opposite trend would be 'risk seeking' (*see* Fig. 3.2).

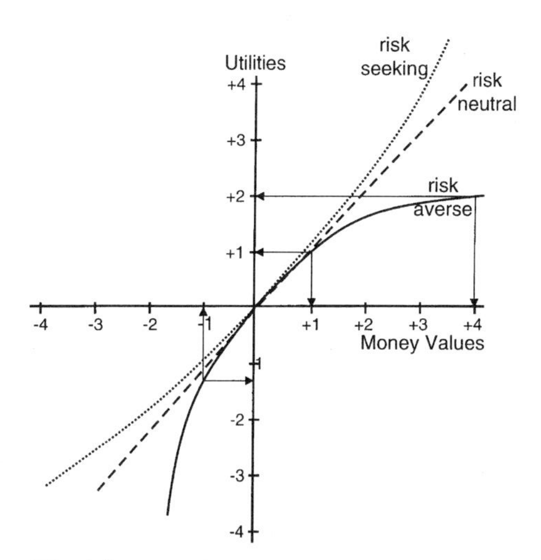

Fig. 3.2 Utility functions of different markers.

I shall not consider here how utility functions are constructed — there are several sources in the literature — but the consequences for decision making are illustrated in Table 3.3, using the initial payoff matrix and the 'risk averse' decision maker from Fig. 3.2. He finds a negative expected utility, and would decide not to drill. Evidently, a 'risk seeker' could choose differently, with, for him, equally acceptable reasons.

Utility functions are specific to individual decision makers (or committees of decision makers in a corporation) and to the circumstances of a particular decision making situation. Such functions often show different trends for low-value, as distinct from high-value, sums. For

Table 3.3 Payoff matrix – the 'utility' of drilling for gas in the back yard

	Payoff	Probability	Utility	Expected utility
Dry hole	(100)	0.6	−1.4	−0.84
Small find	100	0.3	0.8	0.24
Big find	400	0.1	2.0	0.20
		1.0		**−0.40**

example, both individuals and corporations often show risk seeking tendencies when relatively small costs or benefits are involved, but are markedly risk averse in the context of large sums of money.

Utility functions can be constructed for payoff attributes other than money. In practice, few if any companies use utility theory in formal decision making. It is simply not feasible to construct realistic utility functions for corporate decision makers and actual corporate decisions. But the theory does throw some light on the way people react in decision making situations, and it can be instructive to use it, for example, to test alternative ways of presenting a project proposal.

3.4 The value of imperfect information

Suppose we have two estimates of the probability of some event – for example, that a project will generate a particular level of return. If we believe that both estimates are equally accurate, our confidence in the average will be reinforced (*see* case A, Table 3.4). But suppose we have two estimates, which we consider are equally accurate, but which differ significantly, as in case B. The question arises: what should we do in case B? This is quite a complicated issue, which turns on how we value imperfect data.

Table 3.4 Imperfect information – different estimates of uncertain outcomes

	Probabilities	
	Case A	Case B
Estimate 1	0.3	0.1
Estimate 2	0.3	0.5
Average of both estimates	0.3	0.3

Thus, if we know with certainty the probability of an event (as in, for example, spinning an honest coin), there is no point in looking for additional information about its probability, for there is no useful further information. We have only to decide whether to accept or reject the risk. But if we do not know with certainty the chance of success, i.e. the estimates diverge, there may be value in getting additional information.

Clearly, if the divergent estimates are far apart, the value of additional information which increases one's certainty is likely to be higher than when the estimates are fairly close together. However, the mere existence of divergent estimates does not in itself justify risk aversion, nor does it say anything about the cost of getting additional information. As in other cases, we have to decide on whether or not to spend money on getting additional data by comparing the costs and the benefits of doing so.

As a general rule, we may say that the cost of uncertainty is the expected value of the benefits foregone if we reject the project, or the expected value of the losses risked when we accept the project. Usually the cost of getting additional information tends to have a straight-line relationship with the amount (the *quantity*) of information obtained. But the relative value of the information (its *quality*) per increment of expenditure tends to decrease ('10 per cent of the cost buys 90 per cent of the information; the final 10 per cent of information costs 90 per cent of the budget!')

If we plot these relationships, we see that the greatest net gain in expected value occurs at a point *x* on the horizontal axis. To the right of *x*, the marginal gain in value is less than the marginal cost of getting it. So the intercept of *x* with the cost relationship is the most it is worth paying for information (*see* Fig 3.3). Alas, you can calculate where *x* is only from Bayes' theorem, which demands that you have extensive statistical data about consequent probabilities. You may have this from, for example. market research, but it is otherwise rare in project situations. Nonetheless, the relationships help to appreciate the pros and cons of acquiring information for your decisions.

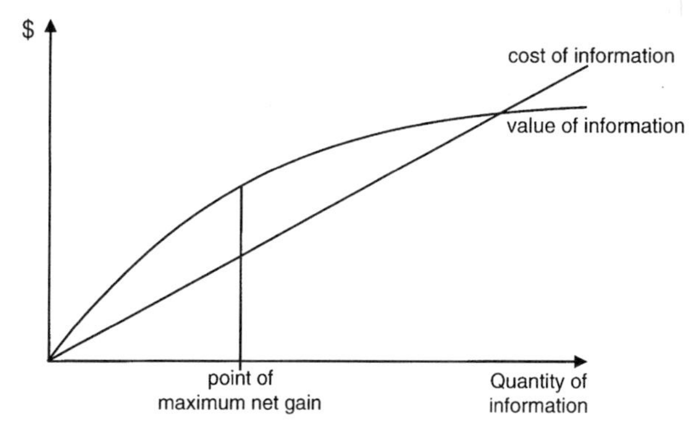

Fig. 3.3 Cost versus value of getting information. To the right of the point of maximum net gain, the marginal benefit in expected value is less than the marginal cost of getting the information.

Large amounts of information make problems seem more complex. There is evidence that, faced with much information, decision makers tend to oversimplify problems. Not surprisingly, the performance of decision makers deteriorates and, although complexity tends to inhibit participative decision making, the involvement of others becomes more desirable.

3.5 Acceptable quality decisions

The purpose of decision making techniques such as those we have discussed above is to reduce the complexity of decision making by introducing a systematic, rigorous procedure, each step of which is accessible to criticism and review. In this way, decision makers hope to make credible decisions and to be able to defend their conclusions, even when the circumstances are hedged around with doubt.

This, however, is not quite enough. Decision making is also a function of leadership – at any rate, in the context of the kind of projects we are discussing – and flawed decisions usually result from a failure of leadership, because those who have to act to implement the decision will either not act or will do so reluctantly. For this reason, we need to consider not only decision making techniques, but the process by which decisions become effective action.

As a project-related generalization, we can say that the *quality* of a decision is related to the three basic criteria by which we judge project success: it meets its performance specifications; it is within its budget; it is within its schedule. Secondly, the decision has to be *acceptable* to those who have to act on it and any others who may be affected by it. The effectiveness of a decision therefore depends on the balance, determined by the current circumstances, required between its quality and its acceptability.

Because acceptability depends on the involvement of other people, we can visualize a number of ways to arrive at a decision based on the degree of other people's participation in the making of the decision:

1. we have all the necessary information, and the competence to use it, so we do not need the participation of anyone else;
2. we don't have all the information or the competence required, so we must obtain help from others;
3. like 2, but we make the decision in consultation with other people;
4. like 3, but the decision is arrived at by compromise over differences, and negotiation;
5. while we have the necessary information, we delegate to others the responsibility for making the decision.

We can see that from 1 to 5 these processes involve an increasing degree of group participation, implying the need for group leadership.

It has been said (Prescott, 1980) that an effective decision optimally combines *quality* and *acceptability*. Quality refers to the achievement of aims, and the cost and time taken to achieve the aims of the decision. Acceptability refers to acceptance by subordinates, superiors, associates and others affected by the decision. However, we may observe the tendency of some leaders to decide on the basis of seeking high quality decisions without taking much notice of acceptability.

The best way to choose an effective decision making process follows from answering 'yes' or 'no' to the following questions:

1. Are decisions likely to be significantly different as regards quality (i.e. in relation to performance, cost, time)?
2. Can I make a high quality decision without the help of anybody else?
3. Do I know what help is needed, and how to get it?
4. Do I need other people to accept my decision in order to get it implemented?
5. Will other people accept my decision if they are not involved in making it?
6. Do the other people share the aims that I want to achieve in making the decision?
7. If I consult them, are the others likely to accept my decision?

Considering the possible answers to each question would allow us to draw a route map of the decision making processes (see Prescott, 1980, Ch. 7). This method is useful when we are trying to ensure the best climate for decision making. As an illustration, try using it to choose an approach to an issue such as this:

> The Minister wants to use half our natural gas reserves to make fertilizer (his family owns a lot of farm land and an important trading company), and has asked me to prepare a study for submission to financial institutions for project funds. How should I go about this?

3.6 Summary

Decisions about projects are unlikely to be wholly objective. They involve personal judgement, even intuition, and the use of information which is in some way incomplete or otherwise imperfect. Early decisions concern the choice of a project concept from among several options, but their influence on subsequent decisions persists throughout the development of the project.

Evaluating projects in order to select the preferred option is not always straightforward. Normally selection will be the outcome of series of decisions made in studies aimed at refining the definition of the project so as to improve the accuracy of estimates of project economics.

A range of techniques is available to support the decision making process. However, quantitative methods do not always take adequate account of the circumstances which influence project decisions. Moreover, because quantification often relies on the use of probabilistic information, the interpretation of project expectations becomes complex and frequently difficult to turn to practical use.

The decisions made have to be acted upon, usually by other people. So the decision making process must be not only credible and defendable, but acceptable to those whose task it is to implement the decisions.

4

Project Risk Management: An Introduction

4.1 Introduction

No-one sets out to sponsor a bad project, yet many projects fail to come up to their sponsors' expectations. The explanation, in general, must be to do with our failure to manage project *risks*. In this chapter, we will not be concerned with strictly formal methods of quantified risk analysis (such as are applied in hazard and operability studies), although it is part of the project manager's responsibility to ensure that these are carried out at the appropriate times. We will be concerned with risk identification, risk management and the perception of risk, because these are part and parcel of project management throughout the project's life.

People are sometimes reluctant to implement risk management because they think the methods are complicated and intimidating, and because they believe that the purpose of the methods is to allocate blame when something goes wrong. While it is true that some – not all – risk management methods are complicated, it is never true that their purpose is to allocate blame. Their purpose is simply to help people to manage projects better.

4.2 Risk awareness

The risks that we are aware of, and plan for, seldom cause trouble. The risks that cause trouble are those that we overlook. Using checklists, one's own and other people's experience, and brainstorming about things that could possibly go wrong both help to highlight the fact that risks are not simply potential problems affecting the physical or commercial security of the project. Risks often arise from different perceptions of the project and different views as to the nature and seriousness of particular situations. A systematic approach to these perceptions and the awareness of risk deriving from them make it easier to plan effective risk management. The method we adopt is, however, less important than the aim: to reduce project risk to acceptable levels.

IDENTIFYING RISKS

It is easy to spend a lot of time on risk identification without achieving much that is useful. Start by stating the aim of the analysis as clearly and concisely as possible, writing down the reasons why it is being done, the nature of the decision(s) required, and the date when the decision(s) are required. This statement of purpose works in much the same way as making an agenda for a meeting. Without it, the meeting may go on indefinitely, clarifying nothing.

Specify the scope of the analysis, e.g. 'This risk analysis covers component B of project "Pyramid" '. Next, write down each of the perceived risks with as much detail as possible. A cause and effect ('fishbone') diagram helps to clarify the relationship between risks and likely outcomes (Fig. 4.1).

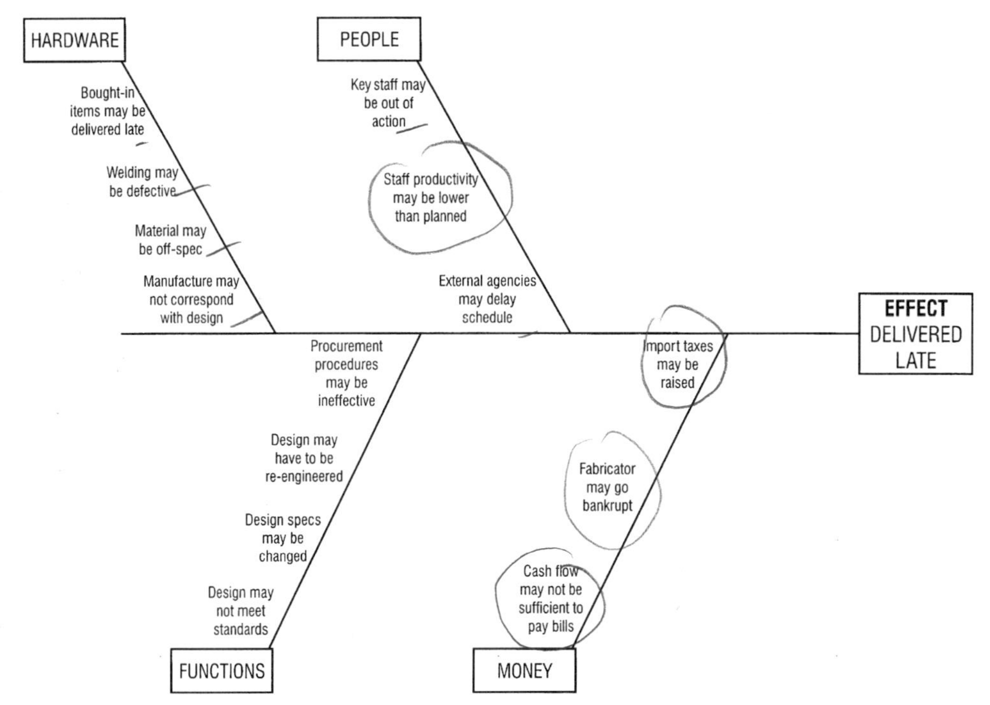

Fig. 4.1 Cause and effect diagram ('component B').

The *effect* can be derived from the statement of purpose, and the *causes* (i.e. risks) can be further broken down into greater detail if necessary. The extent of detail required can be decided by answering the following questions:

- Is the level of detail sufficient for the purpose of the analysis?
- Can the risk be assigned to one identifiable owner?
- Are specific responses indicated?

Negative answers show where greater detail is necessary.

Note that any risk description which refers to time, cost or performance is potentially a statement of *impact*, because these are the measures used for assessing impact.

ASSESSING RISK LIKELIHOOD AND IMPACT

It is usually better to make all the 'likelihood' assessments first, and then to consider the impacts, keeping *likelihood* and *impact* assessments separate for the time being.

Assessing likelihood

Use simple ratings, e.g. HIgh, MEDium, LOw. Ensure that these are used consistently (HI typically occurs more than 50 per cent of the time). Because the selection of ratings is subjective, choices will be biased, usually due to an individual's prior experience. (Aristotle, in the *Poetics*, observed: 'A likely impossibility is always preferable to an unconvincing possibility.') Settling disagreements between individuals by making a compromise is seldom a good idea. It is better to compare prior experience with present circumstances, and choose the more relevant experience as a guide.

Assessing impact

Use the same HI, MED, LO ratings, although in some cases it is useful to have a CRITical rating as well. Impact is normally assessed in terms of cost or duration (time), sometimes on performance, e.g. typically when time and cost are relatively less important:

- Impact is assessed against the whole project, not just the activities being considered.
- Impact is assessed assuming no response is included. This creates a baseline for assessing the effectiveness of possible responses. People often tend to assess impact under the conditions of an assumed response, so giving the risk a lower rating. The risk will then be given a lower priority, and the assumed response may never be never put into effect!

Actual assessment is not usually difficult. Use the following questions to test:

- Is the risk clearly defined?
- Is a variable involved?
- Are different experiences (or biases) involved?

For example, if the risk is 'there may be a human resource shortage', it is necessary to define the risk in more detail: which specific resource, for which activity? This leads to the identification of critical resources. It is then necessary to clarify 'shortage', which may mean different things to different people, due to their various experiences. Finally re-state the risks, which may now have been broken down further to risks associated with each type of resource and activity.

IDENTIFYING LINKAGES

Treating risks as if they are independent of one another can be misleading. Moreover, one's initial assessment of the likelihood of a risk is often lower than is suggested by the risks that influence it. Dependence between risks can be illustrated by an *influence diagram* (Fig. 4.2).

Write down the risks identified, and ask, 'Does risk A influence, or is it influenced by, risk B?' If the answer is 'yes', draw a line linking the two. (The line can be marked '+' if the influence increases the risk, or '–' if it reduces the risk, or it can be drawn heavier for strong influence and lighter for weaker influences).

It is wise to limit the number of risks in a diagram to between about five to 15 so as to avoid the diagram becoming too confusing. If one risk seems to be the focal point for many linkages, check to see if it is sufficiently clearly defined and broken down to a suitable level.

PRIORITIZING RISKS

Combining the ratings assigned to likelihood and impact gives a crude measure of priority. A risk in which HI likelihood is combined with HI impact would obviously merit high priority. Combinations of, say, LO likelihood with HI impact, and HI likelihood with LO impact, are more difficult to prioritize.

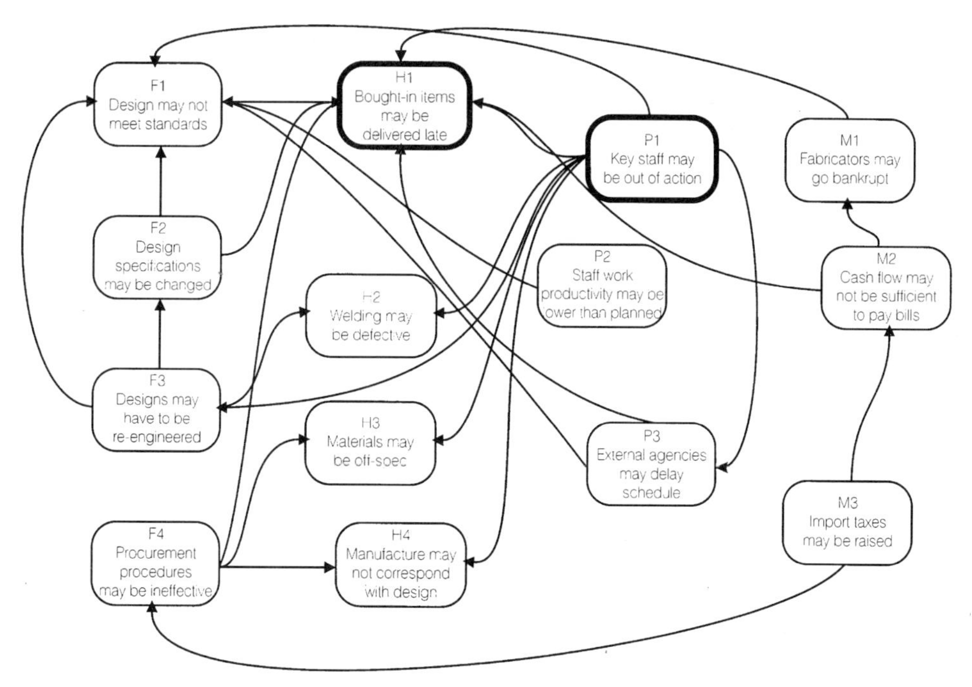

Fig. 4.2 Influence diagram ('component B').

We could rate *likelihood* as a percentage, and *impact* as, for example, days of late delivery (or units of additional cost incurred) (*see* Table 4.1). Fig. 4.3 illustrates this. Which ever method we use, we must be consistent in our use of ratings and cautious with numerical data, for we are not likely to have reliable statistics with which to support real quantification.

Table 4.1 Prioritizing risks: example of how qualitative estimates may be roughly quantified to get an idea of their relative importance

Likelihood	Impact
LO = 0–10%	LO = 1–5 days
MED = 11–20%	MED = 6–15 days
HI > 20%	HI > 15 days

Now consider the influence diagram in Fig. 4.2. It indicates that risks H1 and P1 are the main focal points of influence, either influencing or being influenced by other risks. We should see if these risks can be broken down further by, for example, specifying bought-in items by type and source in H1, and identifying key staff members in P1, then rating the broken-down risks. Lastly, we should consider revising the likelihood ratings to allow for the concentration of influence on these risks.

This procedure produces a prioritized list of risks, but does not provide a cut-off point separating risks which must be attended to from those which could be ignored. Managers claiming vast experience are likely to take the view that some 20 per cent of the risks cause most of the problems, so that only the top 20 per cent of the prioritized risk list is worth worrying

CAUSES, i.e. RISKS

	Likelihood		Impact		Effect
	Rating	Percentage	Rating	Percentage	(expected no. of days late)
Functions					
F1 Design may not meet standards	M	M	20	10	2
F2 Design specifications may be changed	H	L	30	5	1.5
F3 Designs may have to be re-engineered	L	H	5	20	1
F4 Procurement procedures may be ineffective	L	H	10	20	2
Hardware					
H1 Bought-in items may be delivered late	L	L	5	5	0.25
H2 Welding may be defective	M	H	20	15	2.25
H3 Materials may be off-spec	H	L	25	5	1.25
H4 Manufacture may not correspond to design	L	M	5	10	0.5
People					
P1 Key staff may be out of action	L	H	10	20	2
P2 Staff work productivity may be lower than planned	M	M	15	10	1.5
P3 External agencies may delay schedule	L	H	10	30	3
Money					
M1 Fabricators may go bankrupt	L	H	5	40	2
M2 Cash flow may not be sufficient to pay bills	L	M	5	15	0.75
M3 Import taxes may be raised	H	L	30	5	1.5

Fig. 4.3 Component B: effect = late delivery.

about. A better approach is to choose the cut-off point by reference to the acceptable level of residual risk for the project scope under consideration.

RESPONDING TO RISK

A 'response' is any action implemented to deal with a risk or combination of risks. A typical classification of responses was given in Section 3.3, 'Sensitivity and risk' (p. 30).

- avoid
- reduce
- transfer
- accept
- contingency.

Accepting risk means, for example, that we accept to pay for losses; sell assets; raise a loan; or cover by insurance in-house. Reducing and avoiding risk may be achieved by design; by selecting suitable hardware; by using appropriate procedures; and by suitable training. Transferring risks can be achieved *via* contractual arrangements and/or use of outside insurers. There will always remain some residual risk which has to be accepted and which may be covered by contingency allowances.

It is usually best to consider the *timing* of the response first, and worry about the type later. Decide whether the response should be implemented *before* or *after* the risk occurs.

A contingency response (typically an additional allowance of time and/or money that can be used as a general response to a number of risks that are often not specified in detail) involves two actions: one before and one after the risk occurrence, in which the 'before' action enables the 'after' action. In this case, we must define a *trigger* point for implementing the 'after' action.

A method for generating responses for an identified risk is to draw a line horizontally across the middle of a large piece of paper, with 'before' above the line and 'after' below it. Then brainstorm the possible responses to the risk. This highlights the range of options before the risk occurs in comparison with the number after.

One option is always to 'do nothing'. This avoids all risks resulting from a particular response action. Doing nothing ought therefore to be considered among the possible responses (which is not to recommend doing nothing as a general principle!).

Selecting responses

Choosing a suitable response requires appraisal of the effect that implementing the response will have on the original risk. A *cost/benefit* approach is useful: determine the effect that each response has on the original risk ratings, and the cost of implementing the response. The procedure is as follows:

- create a baseline (usually the 'accept' response, i.e., the original risk rating);
- consider each possible response and estimate the cost of implementing the response and the effect it has on the original rating of *likelihood*;
- reassess the *impact* of the risk, assuming it occurs, given that the response has been implemented.

The following should be noted:

- When a response totally avoids a risk, it is necessary to assess the *secondary risks* associated with that response in order to make a useful comparison.

- The *impact* of a risk is often not affected by a response that reduces the *likelihood* of its occurrence.
- Often a single response will be implemented for a number of risks, the benefits of which may not be apparent when evaluating the risks individually. Make a record of these additional benefits.
- If you consider only the 'worst case' impact, you will not be able to differentiate between the potential responses. Use 'best case', 'most likely' and 'worst case' impacts to demonstrate the effectiveness of a response.
- Wanting to implement two or three responses to a risk may make you – if you are a project sponsor – feel good, but usually has little if any cumulative effect on the likelihood or the impact of the risk. It will however cost more.

CREATING A RISK MANAGEMENT STRATEGY

You will need to decide on the way the risks will be reviewed, and how new risks will be identified. Decide also on the level of necessary documentation. Ensure that the results of all assessments are properly documented.

In large projects, appoint someone to be responsible for coordinating and recording all risk management activities. Responsibility for managing the risks nonetheless remains with the respective 'risk owner', and ultimately with the project manager.

Risk management is an integral part of project management.

Project people are often concerned about excessive documentation and bureaucracy, and the time and effort involved in risk management. These concerns can lead to negative attitudes, demotivation and neglect of potential problems. Running risk 'workshops' helps to build project teamwork because they give a positive atmosphere to working on a key issue. People can't be blamed for something that has not yet happened, but people will work together on suggesting how to overcome potential future problems.

Risk assessment does not guarantee correct decisions. However, it does enable people to make better decisions. Risk management should be seen as an investment, not an overhead.

4.3 Risk classification

Classifying risks by type makes it easier to generate appropriate risk responses. The classification can be done in several ways. Often the most accessible way is to use the project work breakdown structure (WBS) as the basis for classifying risks associated with, for example:

- hardware
- functions
- resources
  - financial
  - human
  - matériel
- responsibilities
  - assignment of identified risk 'owners'
  - contractual.

Another approach is to classify risks by the area of the project where they have most impact, e.g.:

- *schedule risk* – based on a breakdown of the project by activities;
- *capital cost risk* – based on a breakdown of the project by the cost of project components;
- *schedule and cost risk* – based on a breakdown of the project by activity durations and their associated material costs and rates;
- *economic feasibility risk* – project lifetime cash flows are broken down into cost components;
- *contractual risks* – based on a breakdown of the project into ownership/responsibility components.

Project financiers sometimes use an alternative classification which highlights their special interests, e.g.:

- *pre-completion risks* – sponsors, investors, contractors, technical, environmental;
- *post-completion risks* – operations and maintenance, raw materials supply, product sales;
- *financial risks* – cost overrun, foreign exchange, interest rates, inflation;
- *country risks* – political, sociocultural;
- *miscellaneous* – *force majeure*, risk allocation, performance bonds.

The Project Management Institute (USA) and the Association of Project Managers (UK) are among professional bodies that have published risk classifications in their respective *Bodies of Knowledge*.

Whatever way it is done, the breakdown structure should be kept as simple as possible at this stage (e.g. five to a maximum of 20 components). Risks tend to overlap the boundaries of classification structures, so some care is called for to ensure that all aspects of a risk are covered and none neglected.

Occurrences which are potential causes of risk are now identified for each component. Where possible, the reasons why each occurrence is a potential risk should be stated, and dependencies indicated. For example, a particular project activity duration may be liable to extend if labour productivity falls because of adverse weather conditions.

The next step is to identify the responses appropriate to the causes of risk. The simple classification described in Section 4.2 (p. 43) may be amplified as follows:

- *remedial action* – intended to be taken after the risk event occurs, i.e. a contingency plan which is to be acted on only if the risk actually occurs, but not before;
- *remedial and precautionary action* – intended to be taken after the event, but with necessary prior precautions, i.e. a contingency plan incorporating some preliminary action (such as purchasing long lead-time spares in case of future urgent need);
- *preventive action* – to mitigate the impact of the risk, i.e. a contingency plan to reduce future costs or damage at the expense of incurring cost now (such as redesigning to simplify future maintenance);
- *prior action* – to eliminate or reduce the likelihood or the impact of a risk event (this is 'pure' preventive action).

It is important to identify all practicable responses for each important risk situation and to identify secondary (and consequent further) risks if the planned response fails.

The identification and classification of risks must be documented. This helps clear thinking; it is useful for project implementation and start-up; it promotes good communication, especially when briefing new project team members; it provides an audit trail; and it captures expertise for future projects.

What we have covered so far is *qualitative* risk analysis, and is appropriate for the phase of project development when the scope of the project is being worked out. During subsequent project definition, it is useful to continue qualitative analysis to clarify risk issues by identifying linkages between risks and responses; deciding which are major and which are minor risks; deciding whether (and how) a response applies to one or to a range of risks; and deciding what specific action(s) to take.

However, at some stage a *quantitative* approach to risk assessment will become necessary. The problems now are:

- deciding when to use quantitative analysis, and to what extent. An important factor in this decision is the organization's risk-aversion or risk-taking culture.
- how to quantify risk in the absence of adequate probability (i.e. statistical frequency) data about the risk events typically encountered in projects (other than for quantitative risk analysis (QRA) applied to physical risks such as failure of equipment components). Lack of data makes it *more* important to assess the sensitivity of the project to a risk event, and to estimate what probability of risk would justify the expenditure needed to eliminate it.
- how to estimate the impact of risks arising from combinations of conditions. It is tempting to rely on QRA methods, because little else is available to the decision maker. But, because QRA relies on historical data, it may lead to aberrant decisions when it is applied to systems or procedures in which combinations of conditions which result in failure may (1) never have occurred before, or (2) not have been anticipated by planners. Combinations of this sort probably arise in project management more often than we expect.

It is arguable that too much quantification attempted too early merely leads to confusion, and that risk assessment should be kept simple. It is also arguable that there exist many decision aids (in the form of computer software) which speed up and simplify the analysis of a wide range of project risk scenarios. The *interpretation* of risk assessments, however, always calls for mature judgement, which is inevitably subjective, but which should be based on sound experience and proper training.

Making judgements about risks implies that (1) we can identify all the risks in the project, and (2) we know how to rank them in an order which reflects their significance in terms of the project's success. Risk identification relies largely on checklisting, which is useful in so far as the checklists:

- are constructed in a way which is relevant and appropriate to our project;
- are complete and exhaustively comprehensive;
- contain no overlapping items (to avoid double counting);
- do not assume parity of experience between the compiler and the user;
- are free from cultural bias.

Few, if any, published checklists satisfy all these requirements. In practice, published checklists must be used with caution as guidelines to be carefully and thoughtfully adapted to our particular situation.

4.4 Public risk perception

In the sense that technology incorporates an economic dimension, people judge technological advance to be worthwhile because it generates the economic surplus that provides them with the luxury of choice. Technological advance depends on innovation. Choice and innovation bring with them the need to evaluate and manage risk – and we must acknowledge that our perceptions of risk are not always entirely rational.

The public perception of risk can have important consequences for technology-based projects, notably in the energy and chemical industries, because these industries seem remote and mysterious to most of the public. Formally, risk is often expressed as the chance of causing some number of deaths or illness in some stated period of time, but risk can also be expressed in terms of economic loss. In general, we are often mainly concerned with *safety*, a broader concept than *risk*, although the first imperative is to reduce risk to acceptable levels.

This leads us to think about some level of risk which the public, as stakeholders rather than shareholders in a project, would consider as acceptable. What people consider as acceptable depends on several factors. For example, some are prepared to accept a high level of risk in activities which they undertake voluntarily and where they feel that they are in control. But few of us are willing to accept risks imposed on us from outside.

Again, people in general tend to overestimate the risks due to major accidents and to underestimate (in comparison with statistical records) risks due to minor accidents. This has led to stringent regulation of potentially hazardous industrial processes (at considerable additional cost), while relatively little attention is paid to risks (such as motoring accidents) which people have come to accept as part of normal daily life.

Thus, regarding industrial operations, the UK Health and Safety Executive has established (1) maximum levels of individual risk which are defined as just acceptable; (2) minimal levels below which further action to reduce risks may not be required. Between these levels, managements are required to reduce risks to levels 'as low as reasonably practicable' (ALARP) – which means in practice to the point where the cost of further measures exceeds the benefit gained. Estimation of these risk levels is based on well-developed methodologies of QRA in which the probability of risks is derived from statistical frequency considerations.

However, applying QRA to the principle of ALARP has doubtful validity when the perception of risk depends on what are essentially subjective, qualitative points of view. What, for example, is the 'value of life'? We need to know this if we are to be able to compare the cost-effectiveness of alternative proposals to reduce risk.

A measure sometimes adopted is 'the value of one statistical life'. This was formerly estimated by calculating the average value of an individual's economic output, but there are obvious difficulties in applying this criterion widely. Another method is to find out (e.g. from opinion polls or public enquiries) what people are willing to pay for a reduction in risk . The results of such surveys vary widely. Not surprisingly, people are willing to accept much higher costs for risk reduction if the money comes from someone else's pocket – such as the Government's – than if it were from their own (even though they know that the Government is actually spending *their* tax money).

In the UK and the USA, acceptable costs for industry to meet *health* regulations are one to two orders of magnitude higher than the costs of meeting *safety* regulations per nominal life saved. Presumably this reflects public concern about chemical and nuclear risks to health, as distinct from the actual safety records of these industries. For example (Roberts, 1995; 1988 money values):

- personal valuation of one statistical life: £20 000–£80 000
- human capital valuation: £0.3–0.5 million
- median of industrial investment costs surveys: £1.6 million.

Public perceptions of risk are largely conditioned by the information provided through the mass media. Disasters are newsworthy; safety is dull. If all we hear about are explosions, fires, poison clouds and environmental catastrophes then we shall remain worried and fearful. We need more balanced reporting: 'Balanced risk assessment should be an integral part of the policy-making process' (attributed to Prime Minister John Major). Certainly, adverse media reports, even if ill-founded, can exert very serious influence on the prospects of a project – and of the people developing or managing it!

The problem is not so much the calculation of risks, but the public perception of them:

- How do we deal with this problem?
- Whose opinions can be trusted?
- How do we deal with the assumption that an unknown risk is necessarily a high risk?

It has been argued (by R M Aicken, see Roberts, 1995) that the definition of risk as the probability of a particular adverse event occurring in a given period of time makes risk a vector quantity in two dimensions, and that this is far too simple a view. Further, the overestimation of major accidents and the underestimation of minor accidents have rational genetic explanations, because a major accident with widespread effects would have more serious consequences for the human gene pool than minor localized accidents. It follows that it would be entirely rational, for example, to spend larger sums of money per statistical life saved reducing the likelihood of a nuclear catastrophe, than to spend smaller sums/life saved reducing the minor effects of radon in homes. Whether you accept it or not, the argument highlights the need for more study of comparative risk assessment and for more careful communication.

4.5 Effective risk decisions

Effective decision making involves consideration of the quality of information and of decision rules, and the acceptability of the decision making process (see Chapter 3).

Quality of information

Consider the methods used by the decision maker in dealing with uncertain information: sensitivity testing, risk perception and assessment, and risk analysis. 'Uncertainty' is a qualitative condition. It is capable of being quantified – sometimes – by using probabilistic methods. (Bertrand Russell is said to have advised his students that 'probability is the most important concept in modern science, especially as nobody has the slightest idea of what it means'.)

Quality of decision rules

Decision rules must be explicit and compatible with corporate or project procedures. They should enable robust decisions to be made. They must be unambiguous and clearly stated.

Classifying and ranking risks demands that we design and use a suitable ranking scale; assign suitable weights to the risk factors considered; choose a suitable method for combining the weighted factors to arrive at an overall rating; and choose appropriate criteria for accepting the residual risks. With complex information to use in decision making, people often simplify too much. We must check to see if there is evidence of this in our risk assessment.

Acceptability of the decision making process

This implies considering the decision process in terms of achievement of purpose; the total cost of achieving the purpose; the time taken to achieve the purpose, and its acceptability to, for example, subordinates, peers, superiors, unions, stakeholders, outside authorities or agencies. The routes to acceptable quality decisions include *autocratic, informed, consultative, negotiation* and *delegation.*

4.6 Recovering from project risk situations

When a project runs into serious risk of failure, special efforts are needed to bring the project back on to a path to recovery. A systematic approach helps to make these special efforts less burdensome.

RECOGNIZING A SERIOUS RISK OF PROJECT FAILURE

Danger signs include missed dates, defined tasks not done and the need for reworking. Because the project team is very close to details, they may miss signs such as these, or they may not acknowledge that these signs indicate impending failure because they are optimistically committed to project success, and mistake optimism for confidence. Or they may simply not be willing to discuss problem signs because they fear failure (cf. the ancient Greeks who were inclined to kill thc messenger who brought bad news!).

When a problem situation is recognized, it is imperative that key people accept the need for change. The problem will not go away unless decisive action is taken.

DIAGNOSING THE PROBLEM

Project audits are vital in recovery situations. *All* problems must be brought out into the open and investigated. Typically the audit will cover the following areas.

Review of performance and management

Management mechanisms and processes should be clear, visible, and make the physical status of the project readily available. The review must disclose the current status of the project: cost, time, deliverables prioritization, etc.

Review of technical content

This is the review of the status of products or promised deliverables in the context of the business objectives of the project. The review must state what has to be delivered, the quality, major risk areas and contractual issues.

Decisions on remedial action

A fundamental decision is whether to continue or to abort the project. This decision must be based on the conclusions of reviews; the likelihood and impact of risks; the cost of recovery in terms of budget, schedule, scope and expectations; and the willingness of management to support recovery.

Recovery planning

If the decision is taken to proceed with recovery, and this decision is backed by management, the plan must address a range of issues, as discussed below.

People issues

The right team must be put in place. This may mean replacing some or all of the current project team; replacing or reinforcing technical functions; contracting out instead of using in-house resources; or redesigning the project structure and organization. Gaining and keeping the commitment of the new team is critical: encourage 'ownership' of the *future* success of the project and avoid dwelling on past apparent failures.

Technical issues

An important aspect of recovery is reduction of technical uncertainty. This can be achieved by increasing the chances of delivering good quality or by changing what is being delivered to that of proven good quality.

In the case of technically complex or 'leading edge' deliverables, more effort may be put into prototyping or simulation, or into replacing 'developmental' with 'proven' applications. Sometimes pressure of time on resolving technical uncertainties ('concurrency') can be relieved by reducing the project scope or changing the phasing of deliverables to later dates.

Contractual issues

If these cause problems or if the proposed solutions to technical questions raise contractual issues, contracts may have to be renegotiated or even cancelled.

Setting priorities

Because there is inevitably too much to do in too little time, priorities must be established, e.g. by identifying those activities which do most to reduce risk and those that will move the project on significantly. For example, when scoping, designing and implementation overlap and cause the problem (as in fast-tracking), then either:

- freeze the scope, then design and implement in line with the frozen scope; or
- stop and hold design until the scope is clearly defined.

It may also be necessary to eliminate non-essential parts of the project or to revise expectations.

Communications

If the project team has lost commitment, they believe that there is no need for radical change, or they are confused about their roles in a recovery situation, communication becomes vital for, for example,

- generating a realistic plan, with realistic costs and reports;
- implementing straightforward and easily maintained report/approval procedures;
- producing and using 'need-to-know' information;
- straightforward and effective change control, fault reporting and cost control.

Recovery actions

The recovery plan must be put into effect decisively. The plan must make clear what is to be achieved, when it is to be achieved and how it is to be achieved. The project manager has to balance what needs to be done to get control of the project and to enable it to progress, against what has to be done to maintain control over the long term. While *getting* control of the project is mainly a matter of damage limitation, *maintaining* control over the long term is more a matter of managing risks and establishing effective monitoring procedures.

The project manager will also have to balance competing demands from different places, e.g. to satisfy the sponsor's management (who need to know what has happened) and to create space for the project to recover and succeed.

4.7 Efficient project risk management

Many formal presentations of risk management methodology do not make it clear that the techniques must not be applied mechanistically. Efficient risk management calls for an understanding of the totality of the project (or part of the project) under consideration – which may be helped by using checklists intelligently – but also calls for creativity, imagination and lateral thinking. Communication is a vital element in ensuring that risk management methods are used to best advantage. In risk management, it is best to formalize communication.

The ultimate responsibility for risk management falls on the project manager. Subordinate responsibility should therefore be allocated or delegated wherever and to whomever the project manager deems most appropriate for *that* project's circumstances.

It follows that the project manager must understand and approve the methods used, and must delegate sufficient authority to the person acting as 'risk analyst' to ensure that the analysis is independent and objective, because the project manager will use the findings of the risk analysis to develop risk management strategies: first, to reduce the likelihood of risks arising, and second, to minimize their impact should risks occur.

Efficient project risk management depends on recognition that proactive and judicious spending of some of the budget (e.g. time and/or cost) *before* any actual adverse event will provide better control than invoking a mitigation plan only when a potential risk has become an actual adverse event.

The basic steps in project risk management are:

- identify the risks;
- record what has been identified, for the purposes of control;
- predict the likelihood of the risks, and how serious their impact may be;
- decide on suitable action;
- implement the decisions;
- monitor the implementation.

Fig. 4.4 illustrates the risk management process.

4.8 Summary

When we propose to undertake anything in the future, we are wise to consider *risk*. The characteristics of a project make this not only a necessary, but also a crucial, feature of project planning. The very prevalence of risks – potential future problems which could jeopardize the project's success – demands a systematic, methodical approach to risk awareness: the identification of risks, the assessment of their likelihood and impact, linkages between risks, prioritization and the development of a suitable response strategy.

For most projects, the perception of risk by the public (as stakeholders in the project) influences the acceptability of company risk strategies. This influence is most rigorously imposed through legislation, but it may also be felt through media reporting of project developments. It has become increasingly important for project decision makers to take account of public perceptions of project risk.

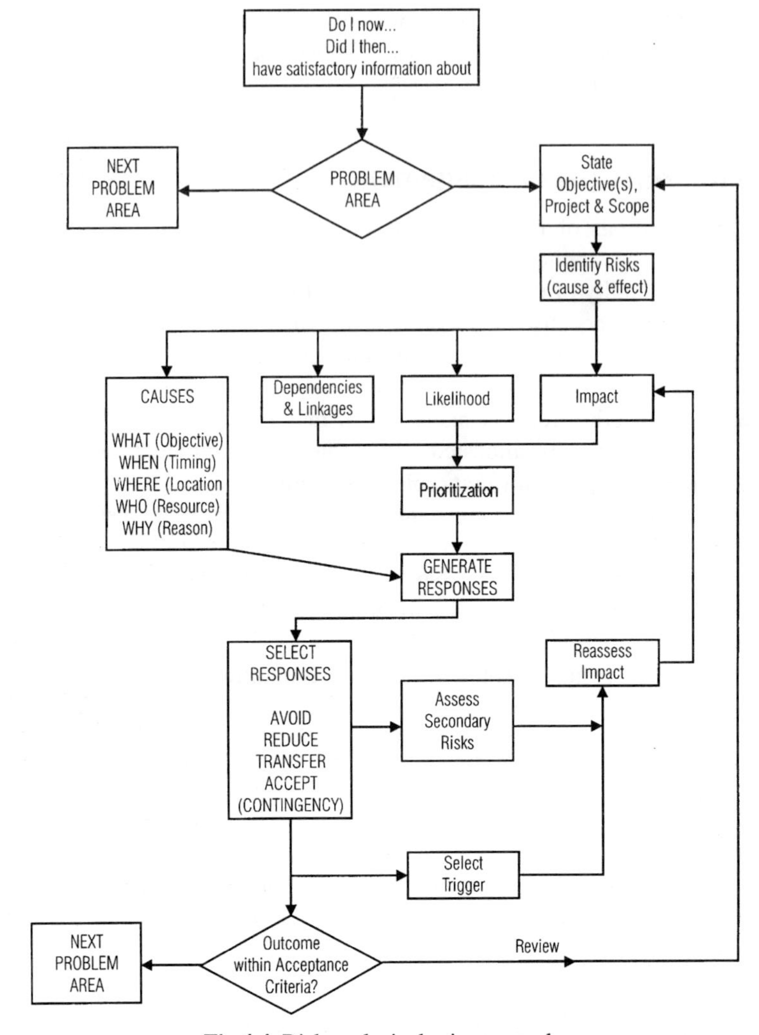

Fig 4.4 Risk analysis: basic approach.

In fact, throughout the life of the project, the communication of decisions about project risk situations is critical in achieving project manageability. In development and implementation, the ultimate responsibility for this – and therefore, for effective risk management – falls to the project manager.

5

The Project Manager's Job

5.1 Introduction

All projects begin as wish lists. The project manager's job is to bring the wishes to reality, safely and fit for purpose, on time and within budget. The project manager may, depending on the scale or complexity of the project, delegate some of his responsibilities to a project team. He remains with overall responsibility for the project as a whole, and it is up to the project manager to decide how he will run his project.

Project leaders have a more difficult task, namely to ensure that the project manager *is able* to manage effectively. Both leaders and managers therefore need to understand what are the project manager's responsibilities, and what authority he should have to act so as to be able to carry out these responsibilities professionally. This chapter aims to provide some guidelines. It may happen, of course, that sometimes *you* are the project leader, project manager and project team all rolled into one. The guidelines will then all relate directly to you!

This chapter is written as though the project we are considering is a typical technical project in which the main players are engineers. However, the concepts of project management responsibility and authority apply to all disciplines.

THE PROJECT TEAM AND PROJECT ORGANIZATION

The project team is the core of the project organization. The functions of the team can be organized in various ways. In any case, a key principle is *single-point responsibility* (SPR). This means that responsibility for specified project tasks is allocated to specified individuals (they become task 'owners'). The structure of responsibilities parallels the work breakdown structure (WBS). Project control rests on this structure.

Project team organization is influenced by the size and complexity of the project, the availability of suitable resources, and the contract strategy if additional resources are needed. Corporate politics also influence project organization. The team structure changes to reflect changing emphases as the project evolves. In terms of its functional composition, however, the team always contains functions such as:

- *engineering* – who provide technical knowhow
- *materials management* – who manage getting the materials needed
- *quality assurance* – who ensure that standards are met
- *project services* – who do the administrative chores.

Additional functions may be called for, e.g. coordination of, and liaison with, third parties; and liaison with operations functions. Other specialist members may join the team for certain periods of time, e.g. spares and maintenance management; construction management; and commissioning specialists.

PROJECT DELIVERABLES

We can say that the project manager is ultimately responsible for 'delivering' the project safely,* on time and within budget, and working according to specification. Similarly, by the same criteria, its component parts are 'deliverables' too. But to deliver the project, or any of its component parts, successfully demands project *strategies* and *plans*. Plans specify what has to be done to bring to bring the strategies to fruition, and how to do it. So the project manager will be concerned to produce a series of plans which are, in a sense, project models of increasing refinement: feasibility studies; development plan; basis for design; project specification; and implementation plan. These plans all contain elements which deal with technical, financial, contracting and purchasing, information flow, and control issues. They should focus on the issues which are critical. Each plan is also a 'deliverable', normally subject to review and approval. Approval, if given, constitutes authorization to proceed to the next stage.

The most important of these deliverables is the project implementation plan, because its approval amounts to the irrecoverable commitment of investment funds. It is, in addition, the main instrument for controlling the delivery of the project itself.

However, if you visit a project manager's office, you will often see hanging on the wall behind his desk the motto: 'Plan the work and work the plan'. He will have his back to it and he will be happy to give it to you as a souvenir when you leave, for he knows that a plan is simply a plausible wish list. Having a plan does not make it happen. You can easily make plans so complicated that no-one can work them. Plans should be kept simple. Another, more favoured, motto is: 'KISS = Keep it simple, stupid!'

TECHNICAL INTEGRITY

Completing the project safely so that it is fit for the purposes of the project objectives implies the concept of *technical integrity*. Technical integrity means that, under the specified operating conditions, there is no foreseeable risk that failure will endanger the safety of people, assets or the environment.

Technical integrity is promoted by using established principles, standards, procedures and work practices throughout the planning, design and implementation of the project, and systematically verifying their application to project deliverables. It follows that any deviation from the approved project plan which is likely to affect the technical integrity of the project calls for immediate assessment and carefully considered approval or rejection. This must be documented to provide an audit trail. The achievement and documentation of technical integrity is another major concern of the project manager.

ACCOUNTABILITY

The project manager is accountable for the performance of his job to his line or corporate

* 'Safely' here should be understood to include the concepts of health, security, safety, environmental and community acceptability which are integrated in the concept of 'technical integrity' discussed in 'Technical integrity' below.

management. But his professional, and primary, accountability for the project is to the 'client', i.e. the project sponsor, owner or user (one party may, of course, wear all three hats!).

Accountability can be thought of as the integration of authority with responsibility. If the project manager has responsibility without appropriate authority, he cannot justifiably be held accountable for the success of his project.

Managers are inclined to take the view that while responsibility can be delegated to subordinates, accountability cannot. A condition for this distinction to work, rather than to be just a semantic amusement, is that work objectives are made very clear, together with their standards of achievement (i.e. their criteria for failure!) in relation to SPR. SPR identifies the task owners. Each task owner is accountable to the project manager for the proper performance of his task work. It follows that the project manager must be able to control the appointment (and, if necessary, the removal) of project task owners.

MULTI-PROJECT MANAGEMENT

Multi-project situations arise when several projects are being managed at the same time within a budget period. When the several projects constitute an integrated approach to implementing a business strategy, managing the integrated project 'system' is sometimes called 'programme management'. As project management methods have become more widely accepted and applied in business development, programme management has acquired something of a cult status. In this section, however, we treat programme management in the same way as multi-project management (at the risk of offending cultists).

The multi-project situation differs from single project management in that multi-project management requires the *relative* merits of the projects within the overall scheme to be taken into account, so as to prioritize the application of resources and the organization of work under varying conditions of resource availability, technical and operational urgency, and economic and project objectives. Any of these factors may suddenly change. If prioritization is not carefully planned, priority tends to go to those projects whose sponsors have the sharpest elbows.

The projects themselves will be of various sizes and in various stages of evolution. It may be that some will be too small to be run by a dedicated team and will consequently use resources on what is essentially a part-time basis. In this case, some people are bound to find themselves having conflicting responsibilities – a conflict which the project manager must resolve.

The aim of the multi-project manager is to manage his portfolio of projects as a whole, phasing expenditures on individual projects carefully so that a specific number of defined milestones, representing a particular level of expenditure, is achieved in a specific time frame. This requires the multi-project time frame to be integrated carefully with corporate budgeting, a procedure which is sometimes difficult to achieve because the programming of projects must be prepared early, when project information is not yet firm. Project programming then often tends to be overoptimistic, e.g.:

- too many projects proposed
- unrealistic milestones
- over-budgeting
- all projects scheduled for earliest timing
- projects scheduled on the assumption of unlimited resources.

Multi-project control mechanisms and documentation are the same as for single projects with the additional requirement that individual project costs, commitments and completion dates must be identified, *plus* multi-project total annual phasing and milestones to be achieved.

Computer-driven precedence networks are suitable for multi-project management, if the system provides:

- a complete overview of the schedule, with the budget features of individual projects;
- corporate budget information separately from project budget information;
- control estimates and latest estimates of the value of work done shown for each individual project;
- for recording the allowances for the various activities;
- yearly phasing and totals;
- relevant performance indicators;
- print-outs which can be used directly for approvals and authorizations.

But remember, computer software cannot make judgements by itself. On issues such as prioritization of work and resources, *you* have to interpret the computer output and *you* have to make the judgement and the decision.

5.2 The project manager's responsibilities

In carrying out his ultimate responsibility for 'delivering' the project, the project manager takes on the responsibility for doing, or ensuring that others do, a variety of tasks. These depend, obviously, on the nature of the project. But they also depend on the phase of the project, so that it would be quite impracticable to try to write down a checklist of all the project manager's responsibilities for all projects. Instead, here is a list of 'prompts' which is intended to help project managers to construct their own 'to do' lists for the project currently in hand.

Throughout the project, the project manager must identify project objectives and their implications for project scope of work and for corporate policy; communicate these objectives to the project team, corporate management and the client organization; organize and motivate the project team to achieve project objectives; and satisfy the client organization. The project manager is the

Fig. 5.1 Shetland – aerial view of Calback Ness before construction of Sullom Voe Terminal. Photograph courtesy of British Petroleum.

prime contact with the client organization, whether the client is internal or external, and as such is responsible for maintaining good, constructive relations with the client organization.

IDENTIFICATION AND FEASIBILITY

1. **Be aware of the results of screening studies.** These studies explain the reasons why the project exists. The evidence is in, for example, economic analyses; studies of layout and major equipment requirements; reserves, raw materials and product studies; environmental impact studies; and the alternatives that have been considered.
2. **Make sure you thoroughly understand the project philosophy** – for example, the objectives of the project (such as emphasis on minimum investment, low manpower, early completion, flexibility), production commitments, penalty situations, investment *versus* operating costs, utilization, maintenance. Pay special attention to the proposed scope of project work.
3. **Ensure preparation of the design basis, project philosophy, milestone schedule.** Although preliminary at this stage, these are the foundations of the development of the project.
4. **Review the design basis and site data.** The design basis, developed from the studies which led to selection of the preferred project option, provides the initial specification of the project workscope, its economic basis, its size and location, and its schedule.
5. **Develop a plan for work on the 'definition' phase.** Prepare and agree procedures, budget and schedule; ensure availability of manpower and other necessary resources.
6. **Establish reporting procedures** for keeping corporate management, client organizations etc. fully informed about project progress. Ensure reports are properly structured and organized, and submitted regularly.

DEFINITION

1. **Develop the project implementation plan.** This incorporates the basic design decisions which affect cost, schedule, safety and quality, and the eventual operational performance of the project. It also establishes the project organization structure; the responsibilities of key people; the critical issues and recommendations for dealing with them; and the contracting strategy. Finally, the implementation plan should contain enough project background information for the project team to understand project interfaces, both within and outside the project organization.
2. **Ensure reviews of design specifications** by operations, maintenance, environmental and safety departments. Review the following with the functional specialists concerned: equipment and material specifications, civil engineering and structural proposals, construction philosophy, pipe routing and equipment locations, tie-in locations, and unusual features of design that may affect cost, schedule or safety.
3. **Identify the operational, maintenance and safety requirements specific to the project,** to avoid having to re-engineer later.
4. **Develop the plot plan.** This establishes the required plot size, the project battery limits and preliminary locations for equipment, taking into account access, maintenance and safety requirements.
5. **Develop the implementation schedule,** based on the implementation plan. The basic components are process design, mechanical design, contract activities, detail engineering, purchasing, construction, commissioning and handover/close-out.

6. **Arrange for preparation of control estimates.** Provide the front-end engineering designs, plot plans and layouts, and specify the class of estimate required. Confirm the overall project scope and the project schedule. Advise on the development of estimates for major equipment items, and in establishing work breakdown and cost accounting procedures.
7. **Ensure availability of project funding,** establishing request/approval procedures if no suitable corporate procedures exist. Pay attention to advance funding of early engineering or purchasing requirements.
8. **Develop a contracting strategy** – prepare a contracting plan and schedule. Screen contractors, prepare and recommend the short-list, prepare and issue ITBs, evaluate bids and recommend award of contract(s).
9. **Establish availability of manpower and other resources** for project implementation.

This work will involve the project manager in a lot of negotiation with people who may be quite unfamiliar with the idea of project manageability. But it is these negotiations which will determine whether or not the project can be managed successfully.

IMPLEMENTATION

'Well begun is half done', according to the old adage. The first day of any project is crucial to getting the project properly launched. Assuming that you have made thorough preparations in the days, weeks, months – even years – leading up to project launch, you are still going to be very busy on day 1. You should:

- confirm pre-contract discussions
- decide on the project team's location, equipment and facilities
- mobilize key members of the project team
- issue project cost codes
- set up communication links
- prepare your 2-week schedule:
  - client contractor kick-off
  - project kick-off
  - task force mobilization
  - print/distribute basic project documentation
  - issue job scope and procedure
  - issue home office budgets: manhours and costs
  - issue initial job controls
  - issue trend procedure
- commence 30-60-90 day schedules
- issue project organization structure
- issue initial control estimate and proposal schedule.

Following project launch, you will be monitoring and controlling progress day by day. The following 'prompts' relate to more or less ongoing activities, with the emphasis changing as the project progresses through implementation towards completion and handover:

1. **Initiate engineering:** explain project workscope to designers; ensure designers have all necessary information.
2. **Periodically review engineering details** using inputs from process, maintenance, safety and functional specialists.

3. **Arrange for support from specialist departments** for preparation of specifications, selection of vendors, invitations to bid (ITBs), bid evaluation and purchase of critical equipment and materials.
4. **Define tie-ins, turnaround and start-up requirements.** Ensure these are incorporated in the design.
5. **Ensure equipment and materials are purchased from acceptable vendors,** review bid documentation, purchase terms.
6. **Ensure proper inspection and expediting of equipment and materials,** especially critical and long-lead time items.
7. **Establish and apply change control procedures.**
8. **Establish and apply cost control procedures,** including cost accounting and reporting.
9. **Prepare construction plan** with advice from construction specialists; construction work packages and all necessary documentation for construction ITBs.
10. **Set up pre-construction review** to ensure a smooth start to construction. Arrange site facilities, services, accommodation etc. for the site project team.
11. **Ensure conformity with construction safety procedures,** including accident reporting and good housekeeping. Verify conformity with design drawings and specifications. Ensure necessary inspection and testing. Monitor productivity, analyse cost and schedule reports, taking corrective action when necessary, control site change orders and expedite deliveries to site. Prepare start-up procedures.
12. **Administer contract(s);** deal with claims, approve invoices for payment to contractors.
13. **At start-up,** check all sections of the project to confirm they are complete and conform to specifications. Arrange pre-start-up review with process design, operations, maintenance, safety and construction specialists to ensure all necessary testing has been completed. Determine mechanical completion and criteria for performance tests. After commissioning and performance testing, prepare a list of 'exceptions' in view of handover.
14. **Prepare documentation for project acceptance and handover, complete outstanding project work.** Complete acceptance/handover formalities. Update drawings to show 'as-built'; account for and dispose of leftover materials; issue final cost report. (Chapter 7 has more about *handover and close-out* procedures).
15. **Write project evaluation report,** including cost and schedule reconciliation, equipment and contractor performance assessments, noting problem areas and decisions/actions taken which may be relevant for future projects ('would do again'; 'would do again, but with some modification'; 'would not do again'), conclusions and recommendations.
16. **Organize project files** for archiving or scrapping. Complete project close-out formalities.
17. **Celebrate.**

5.3 The project manager's authority

In order to carry out his responsibilities, the project manager is delegated *authority*. There is often confusion or conflict in the corporate organization about the extent of the project manager's authority. Corporate confusion and conflict work against the successful completion of projects. The best way to avoid this is to write down the *terms of reference* for the project manager, and ensure that the terms of reference:

- are agreed by the project manager and approved by the chief officer(s) of the organization(s) which have dominant interest in the project;
- designate the project manager by name;

Fig. 5.2 Sullom Voe terminal, Calback Ness, Shetland – aerial view of the process area, looking east (October 1982). Photograph courtesy of British Petroleum.

- specify the organizations which may be involved with the project (especially those from which the project manager is expected to receive support), the interface relationships between them, and the lines of communication with them;
- define the working relationships between the project, other corporate projects and corporate service and support groups;
- specify the project's priority;
- specify the project scope of work, detailing the responsibilities of the project manager in relation to the project scope of work;
- authorize the project manager to control allocation and use of all resources including finance approved for the project, specifying clearly any limits to his authority;
- specify any special deviations or exemptions from corporate policy which apply to the project manager's performance of his duties;
- specify the location and facilities of the project office and any administrative support facilities available;
- define the organization of the project team, specifying adequate manpower for both the project team and any liaison functions under the control of the project manager;
- specify procedures for recruitment to the project team, dismissal from the team and dispersal on project completion;
- define who is responsible for public/media relations;
- are current.

The project manager will be able to exercise his authority more effectively if his corporate standing is high enough for him to be accepted as the agent of the parent company when dealing with other organizations. His authority will be more soundly based if corporate management invite, and pay attention to, his opinions on:

- the project objectives and their likely repercussions on the corporation's overall business;
- features of the project involving unusual complexity or technical uncertainty;
- features involving unusual interface relationships (interdepartmental, intercompany, national, international);
- features involving unusually difficult operations or timing;
- features involving multi-project management issues.

5.4 Project management practice

The project manager's involvement with his project begins – or should begin – when the project starts. It continues – or should continue – until the project ends. There is a further activity in which the project manager should be involved, namely, post-project evaluation. This normally takes place some time after completion of the project. It is intended to assess how well or badly the project has been carried out so as to provide lessons from which we may learn how to carry out futurc projects better (see Chapter 17).

In some organizations, the project manager is seen as being mainly concerned with the implementation phase. In these organizations he may not be assigned to his project until the beginning of the implementation phase. He will therefore have little or no knowledge of important decisions taken in the earlier phases. In the extreme, he may be assigned to manage a project which is simply unmanageable.

A project is more likely to be successful if the project manager is appointed as soon as a real project exists, i.e. at the feasibility phase. This presupposes that the project manager understands the methods of developing projects as well as the methods of implementing them.

However, the project manager's role in each of the phases of project evolution is often unclear. In some situations he may never have full leadership except within the boundaries of his project. If some sort of corporate benefit is thought to flow from arrangements of this kind, it must be weighed against the problems of demarcation of authority that will certainly ensue. It is simpler, and more effective, to involve the project manager in feasibility studies as an adviser. He will assume leadership when the project enters the definition phase and should retain this role throughout the project until final close-out.

PROJECT OBJECTIVES AND WORK CONTENT

In order to achieve the project objectives, work has to be done. The project evolves through several stages. The objectives and the work content of each stage are as follows:

- **Feasibility** has two objectives: (1) to identify the project as a business opportunity; and (2) to assess the project's technical and commercial feasibility. The work content of *identification* is simply to investigate the concept and provide evidence to support further work in the form of feasibility studies.

  The objective of *feasibility studies* is limited to providing evidence to support the funding of work on project definition. The work content of the feasibility studies is to determine whether or not a realistic project actually exists and, if it does, what it consists of.
- **Definition** has three stages: (1) planning project development; (2) preparing the basis for design; and (3) planning project implementation. The objective of *planning project development* is to work towards a formal specification of the project which is capable of being accepted in the corporate programme of business activity, with a budget sufficiently firm that it is capable of receiving approval. This requires detailed economic appraisal of project

options and risks, and assessment of potential physical hazards, to the level where the project development plan can, if legislation so requires, be submitted as a formal document to government authorities.

The objective of the *basis for design* is to define the scope of project work, and the philosophies, procedures, codes and corporate or professional practices which constitute the design standards for project deliverables. The objective of the *planning of project implementation* is to develop a plan for realizing the specified project which is capable of final approval. This represents irrevocable corporate commitment to the project. It also establishes the datum against which all subsequent work, including the performance of the project manager, will be measured.

The overall work content of the definition phase is to progress the project to the point where detailed design can begin.

- **Implementation** has four stages: (1) detailed design; (2) procurement of equipment and materials; (3) construction; and (4) completion. The objectives are simply to complete the work content of each stage as planned. The final stage includes commissioning, initial operation and handover, and project close-out.

To summarize:

1. The objective of each stage is to be able to proceed rationally to the next stage.
2. The work content of each stage is concerned with increasing the definition of project scope, functionality, cost estimates and project schedules.

PROJECT STRATEGIES

The project manager's first step is to formulate a strategy for the planning and control of work. In fact, he must develop – or cause to be developed (for in all but the smallest projects he will work with a project team) – several strategies in parallel:

- **functional** – e.g. the philosophy of design and eventual operation of the project; quality assurance, control, improvement and management; health, safety and environmental/ecological strategies.
- **organizational** – e.g. the resources required and their availability to the project; the division of work; the possible use of consultants and/or contractors; licensing, contracting and purchasing strategies; home office and work-site logistics; how to manage the interfaces between the various parties engaged in the project; communication and reporting.
- **financial** – e.g. how to ensure the availability of funds in the amounts and at the times demanded by project commitments so as to ensure smooth progress; monitoring and control of commitments and expenditures.
- **overall –** the strategy of prioritization: what has to be done first, and in which circumstances.

These strategies guide the carrying out of all work throughout the project.

PROJECT MANAGEMENT TOOLS

Project management methods are basically of three kinds: those which specify what has to be done; those which specify who is responsible for getting the work done; and those for controlling project work so that the project manager can *deliver* the project successfully. Practitioners use a range of procedures and techniques to assist methodical project management. They constitute a kind of 'toolkit' for project managers, and while project managers must understand how to employ them, everyone involved in a project should appreciate what they can – and, just as important, what they cannot – be used for.

The 'tools' and their uses are described in the following chapters. Several of the 'tools' are useful in more than one area of management, and appear in more than one chapter. However, the main references are:

- **Specifying the work to be done** in relation to costs (Chapter 8) and assessing profitability (Chapter 9); in relation to time (Chapter 13); in relation to quality (Chapter 14); and in relation to health, safety and environmental management (Chapter 15).
- **Specifying resources and who does the work** in relation to organization (Chapter 6); financing (Chapter 10); contracting (Chapters 11 and 12); and reporting and information systems (Chapter 7).
- **Controlling the work** in relation to budgets (Chapter 8); commitment control and control of change (Chapter 13); purchasing control (Chapter 14); reviews and approvals (Chapters 7 and 16)

5.5 Summary

In reviewing the project manager's job, we have looked at practice and methods from the standpoint of a project where the magnitude and complexity of the management task justify extensive and detailed control, and where task ownership is primarily assigned to technical functions. This almost inevitably leads us to emphasize the importance of *procedures* for management and control.

However, it could be argued that this emphatic reliance on 'proceduralization' is not justified in smaller projects. It is an argument with some merit if we're talking about the blanket application of heavily documented procedures to control, say, an office move for 10 people. But this is not the real issue. The real issue is that, whatever the size and type of our project, we – leaders and managers alike – have to consider which of the tools in our toolkit we intend to use, and how we intend to use them, to help make our project a success. The choice is ours, and as professionals it is up to us to make it conscientiously, from a sound knowledge of what is possible and what is practicable.

6

Project Organization and Control Principles

6.1 Introduction

An organization consists of the people who are engaged in its activities and the structures according to which their efforts are directed and controlled. Organizations are not all alike. People are not identical and organizational structures cannot be considered as 'standard'. The kind of organization which works well for termites (whose projects are bigger, in scale, than anything *homo sapiens* has managed) is not likely to suit us.

Ideally, our project organization should be compatible with the size and experience of our company, its technology and the products it delivers, and its industrial, market and community circumstances. Further, the ideal organization should reflect what we might call 'cultural' factors: the company's historical background, its complexity and stability, and the way power has been, and is, exercised – both internally and externally – to regulate it.

Within this broad framework, however, we can identify a few basic components of any organization:

- the people who guide the work:
  - overall, long term, strategic: the *top management*
  - day to day, short term, tactical: the *middle management*;
- the people who do the work: the *workforce*;
- the people who plan and design systems for operations and control: the *'technicians'*;
- the people who provide administrative and other support functions: the *support staff*;
- lines of communication, reporting and responsibility.

Not all organizations need all these components, but the way the components are put together determines the structure of the organization (*cf.* Section 5.1).

Probably the simplest case is where the workforce is directly supervised by top management (with minimal middle management). Examples are a company start-up or a small family-owned company. On the other hand, the organization may develop through standardization of skills (perfecting rather than innovating), when the workforce and its support staff will tend to dominate. An alternative is development through standardization of output (as in the product lines of a big company), when the organization tends to rely mainly on its middle management.

Finally, we can visualize an organization in which middle management, the 'technicians', the workforce and the support staff are amalgamated in a mix adapted to achieve a specific, time – and cost-limited, objective. This is a *project team*.

What is likely to cause such a special structure to be preferred, bearing in mind that it will disrupt the normal line-and-staff power hierarchy in the company? If an operating company decides to embark on a new capital development, it may well decide that it is necessary to create a separate organization, outside the existing departmental structure, with the sole responsibility for carrying through the new development. The reason for this is the large number of different agencies which will be involved, all of whose inputs will have to be coordinated if a successful outcome is to result, and the wish to avoid disruption of the company's normal business management activities.

The agencies likely to be involved in a new capital development scheme are:

- various departments within the owner organization
- contractors employed on the project
- suppliers of materials and equipment
- government and local authorities
- 'certifying' agencies
- the operating organization created to run the new facility.

The most important issue is to provide effective management of the relationships (*the interfaces*) between the many different elements which have to be brought together in creating the new development. This is a key function of project-based management.

6.2 Overall project organization

TYPES OF OVERALL PROJECT ORGANIZATION

Four types of overall project organization can readily be identified (*see* Fig. 6.1):

1. **Owner management** – overall project management is provided by the owner. Contractors are employed to undertake the detailed design, procurement and construction.
2. **The integrated project team** – the owner and contractor form a joint team to manage the project. This gives the owner more control over details, and he will normally appoint the project manager and other senior positions in the project team. The integrated project team

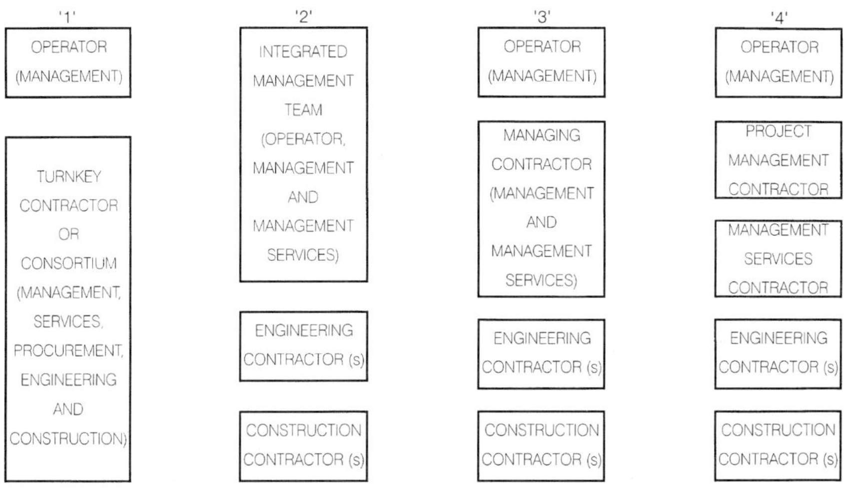

Fig 6.1 Typical project organizations.

provides overall management. Contractors are employed for detailed implementation work as in 1.

3. **The turnkey project** – a single contractor or consortium of contractors undertakes responsibility for the complete development, e.g. management, design, procurement and construction.
4. **The managing contractor** – a contractor is responsible for overall project control and other supporting administrative services. (Alternatively, a separate contractor may be appointed to provide these administrative services.) The managing contractor will supervise other contractors engaged to undertake design, procurement and construction.

On a very large capital development, different organizational arrangements may be adopted for various parts of the project, depending on circumstances. For example, a section of the project may be sufficiently self-contained to lend itself to a separate 'turnkey' approach, while other sections are run by the owner or managing contractors.

CHOICE OF OVERALL ORGANIZATION

Because each project is unique, the project organization should also be unique, i.e. specially adapted to the particular development. The temptation to try to use a standard off-the-shelf organization must be resisted. The factors which can affect your choice include:

1. **experience** – what has been, or can be, learned from the organization of previous projects.
2. **resources available** in the owner organization. The extent to which contractors are employed will depend on the extent to which the owner's in-house technical resources are suitable and available at the time they are needed.
3. **degree of control** to be exercised over consultants and contractors. This is influenced by the complexity and innovative features in the new development. It also depends on corporate policy regarding control of new developments.
4. **contracting and purchasing policy,** which may be constrained by corporate policies which will impact on the type of organization that can be adopted.
5. **geographical location and environment** in which the new development is to be built, i.e. to build a new pipeline system which extends over hundreds of kilometres requires a different organization from that required to develop a new petrochemical project on a single site.
6. **extensions to an existing plant** will require a different organization from that required for a 'grass roots' development.

6.3 Project team organization

Companies sometimes try to run projects by assigning people from the corporate organization to work on a project while they remain in their normal functions within the company hierarchy. In effect, they become a part-time project team. Except for very small projects, this is a penny-pinching approach which commonly runs into difficulty under the pressure of project work – even on small projects. So we can say that practical project team organization falls into two broad categories: (1) *matrix* organization, and (2) *task force* organization.

MATRIX ORGANIZATION

In this case, functional responsibilities are separated from the overall project responsibilities for time and cost control, e.g. a central drawing office may be used to carry out the design work, which may not be the only work that the drawing office is doing at the time. Similarly, on a very large site, where a number of individual developments are proceeding at the same time, it may be convenient to use an existing construction organization to control all construction work.

The advantages of the matrix approach are its flexibility in the use of resources and the more efficient use of experience and specialist services. A central planning department can, for example, service a number of projects. Where an operating organization is undertaking a series of small developments, it may be convenient to assign a small project team to coordinate the work of both the internal departments and the contractors involved.

The disadvantages are the potential for poor communications and control. Conflicts between authorities and responsibilities can arise. The project manager may well have difficulty in securing the necessary priority for his project to meet his programme commitments.

These problems – of communication, control, authority and responsibility – are essentially problems of *reporting*. The matrix organization for projects requires people to report to at least two superiors: their functional superior for general administrative purposes and their project superior for the purposes of the project. It may be thought that this should ensure a more balanced approach to decision making under pressure. In practice, it often leads to confusion and causes delay at project interfaces. Not surprisingly, people are reluctant to change reporting lines which have worked well in the past (i.e. before the project came along). Because project decisions are often taken under pressure, they may be delayed by filibustering for compromise between the alternative reporting lines, or deferred by adoption of ostrich management techniques (if you keep your head in the sand long enough, the problem will go away!).

These difficulties will be most severe when the person assigned to manage the project (or the programme) has relatively low status in the company and lacks support from more senior people.

TASK FORCE ORGANIZATION

The project team is built up as a self-contained unit having access to all the skills necessary to carry through the complete development. The design, procurement and construction managements all report to the project manager, as do those responsible for project services such as cost control and planning. There are considerable advantages if the whole team can be located together in a single location (*see* Fig. 6.2).

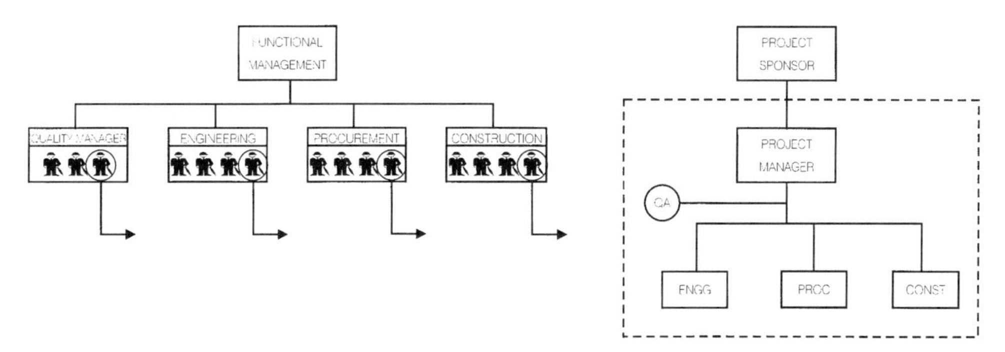

Fig. 6.2 Building a project task force.

The advantages (to the project) of the task force approach are immediately apparent. Team spirit and loyalty to the project are more easily achieved, leading to a greater drive to meet project objectives. Authorities and responsibilities can be more clearly defined, and the dependency of one function on another is more obvious. Lines of communication are direct, and if the team is in a single location, communication will be further improved.

The task force approach does, however, have some disadvantages. It is less flexible in the use of specialist staff, and can promote the inefficient use of resources. Career opportunities within a task force are often perceived to be limited. There is sometimes a tendency for a large task force to ignore corporate policy in the interests of project expediency, and become a 'barony' in its own right. Firm guidelines regarding the limits of authority of the project manager are essential. But note that these limits impose obligations on corporate management as well as on the project manager.

In practice, large project organizations usually evolve into a combination of matrix and task force types. Where the owner organization has a large technical department with specialist divisions, it is usual to call on these for advice as and when it is required, rather than to assign a specialist permanently to the project team where his services may not be fully utilized.

Engineering contractors usually establish a task force on being awarded a major project contract. Owner companies usually rely on the matrix approach except for very large projects.

STRUCTURING THE PROJECT TEAM

Any project passes through a number of phases, such as feasibility studies; definition; detailed design; procurement of equipment and materials; construction; commissioning; handover.

Study work is often considered a 'pre-project' activity, so that the preparation of 'basis of design' documentation may be the first activity for which the project team – as a team – is responsible. In some cases this documentation will be prepared within the owner's existing organization by the technical and operating departments. Alternatively, a separate engineering contract may be awarded for it. The remaining phases will be the clear responsibility of the project team.

Detailed design, procurement, construction and commissioning phases often overlap (e.g. the start of procurement may precede completion of detailed design), but each phase presents its own organizational problems. Consequently, the project organization must be dynamic, for it is subject to change, and it will be necessary to adapt the organization of the project team to suit the changing circumstances throughout of the life of the project.

SELECTING A PROJECT TEAM

Several studies have been published which attempt to predict the composition of a successful management team in terms of 'types' of team member and the circumstances of the particular management task (e.g. Reddin 1987). However, the success of a team typically depends on recognizing that there are relatively few useful roles in a team, and on ensuring that (1) interrelationships between roles are well managed; and (2) individual roles are well performed. In this view, a successful team will have the following characteristics:

1. The team leader is an authoritative but patient figure who generates trust, and looks for and knows how to use ability.
2. The team contains one very clever member, who provides creativity.
3. The team as a whole has a wide spread of intellectual ability. Such teams work together better than teams which are intellectually more homogenous.
4. The team should have a fairly wide spread of personality attributes. (i.e. a member who is conservative, dutiful, predictable; a member who is painstaking, orderly, conscientious; a member who is extrovert, enthusiastic, curious, communicative).

5. The team members' attributes should be matched to their responsibilities and duties in the team, rather than on, say, the basis of their past experience. Some functions may be shared by flexible pairing.
6. Team members should be able to recognize their own areas of weakness, and be prepared to do something about it.

We could visualize that the ideal team composition should be chosen to suit the nature of the particular project which we are to manage. For example, a somewhat different team mix would be needed for a research/development project compared to, say, a conventional, but fast-track, project (Belbin, 1981, 1993). But clearly it will not always be possible to select the ideal team from the resources available, and of course team-building formulas do not guarantee a successful team. They may, however, help to focus your thoughts when you have to build a team for a particular project.

MOTIVATING A PROJECT TEAM

Given that corporate management have not completely frustrated your attempts to select a project team of normal people, you have to work with it, which – if you are the project manager – means that you have to motivate the team members (or, to be more precise, to reinforce their self-motivation). How to do this? Guidelines on team motivation are usually, perhaps inevitably, in the class of 'motherhood and apple pie': they are by definition 'good and beyond criticism'. So are those that follow. At least they are brief:

- keep a positive attitude, even when things are not going well;
- the team members are *professionals* (after all, you chose them) and they will do a good job if you show that you have confidence in them;
- listen to, support and help all team members;
- ensure good informal contact (e.g. daily) within the team;
- ensure good planning so that everyone knows what to do, how and when to do it, and is working together;
- make sure that promises given about plans, schedules and costs are kept;
- minimize bureaucracy: don't relax that minimum;
- deal with team problems and complaints quickly, within the team, even if this is difficult or painful: remember there is room for praise as well as criticism.

PROJECT TEAM FUNCTIONS

A typical project team provides the following functions:

- project engineering
- procurement
- project services
- construction
- quality assurance.

Project engineering

This function is concerned primarily with the development of the detailed designs for the project and the transmission to site of all the necessary drawings and data required for the construction.

Project engineers manage the development of specific sections of project work, and are responsible for ensuring that their sections are completed to schedule and within the budget. In some cases, an engineering manager may be appointed to control the actual detailed design work, but the project engineer will retain schedule and budget responsibility.

Procurement

This function is responsible for obtaining the equipment and materials required for the project and for getting them to the project site. Inefficient control of procurement is one of the most frequent causes of delay and additional cost on a project. Interfaces with the design function are complex, and procurement activity justifies special management attention. Procurement management is covered more fully in Section 14.3.

Project services

This includes the functions of planning (scheduling) and cost control. Sometimes the procurement function is also treated as a project service, because procurement interfaces with both planning and cost control.

The scheduling and cost control functions are very closely connected. Costs are directly linked to progress, and delays invariably involve extra cost. It is quite usual for computer systems to be used in order to correlate and monitor both functions. This does not, however, absolve the project manager from his responsibility for managing the project!

The close relationship between planning and cost control is not always appreciated. On large projects it makes sense to appoint a project services manager who has overall responsibility for both.

Construction

Construction management's main responsibility is the control of all site work. The construction management team is built up as the site is developed. However, the key construction management positions should be filled when the project is in its early stages. Construction management can then be involved in the overall planning and estimating, in the development of the contract strategy, and in the choice of types of contract and of construction contractors, all of which have an important bearing on the success of the construction phase of project work.

Quality assurance

Quality assurance (QA) should be quite independent of all other project functions, reporting directly to the project manager. QA is involved in all aspects and phases of the development, and interfaces with all the other management functions.

There is often a misconception, even by project management, that QA is primarily an inspection function. It is far wider than that. It involves formulating and implementing definite policy regarding the aspects of quality that are essential to ensure that the project meets the criteria of fitness for purpose and technical integrity. QA has to ensure that the agreed policy is enforced through all stages of the project, from initial conception to completion of construction. Quality assurance is discussed in Section 14.2.

6.4 Organizational aspects of project control

DIVISION INTO MANAGEABLE PARTS

A large project can be subdivided into distinct parts. For example, a gas transmission development could be divided into the pipelines, the pumping and compressor stations, and the gas treatment plant. A process plant could be divided into the individual process units and the supporting utilities required to service them. In this way, analogous to a work breakdown structure (WBS), the organization for the largest and most complex project can be divided into separate manageable parts. This is the first step in organizing effective project control.

When a contractor is employed, owners often set up a parallel project team to monitor the contractor's performance. The monitoring team need not, of course, be as large as the contractor's team, but it should have at least the same measure of functional competence.

Because owners often retain responsibility for site work while assigning all other work to contractors, control of construction work at site obliges the owner to set up its own special group, which will be primarily concerned with inspection, monitoring and checking progress, and with ensuring that specifications are properly met by the contractors.

INTERFACE MANAGEMENT

In planning the organization of a project, an issue of some difficulty is planning for 'interface' management. Indeed, project management as a whole has been defined as the management of all interfaces, classified as:

- personal interfaces (i.e. requiring definition of individual responsibilities and authority);
- organizational interfaces (i.e. requiring definition of the scope of activity of organizational groups or activity cells);
- system interfaces (i.e. interfaces between the physical project and the rest of the physical world).

Interfaces follow from the subdivision of project work activities (e.g. in a WBS) and the consequent allocation of specific responsibilities. An interface demarcates the source of information from its intended destination and use. Interfaces are potential barriers to communication and the exercise of responsibility.

Interface management is an important aspect of the control of project changes. This resembles the operation of an instrument loop in control theory (Fig. 6.3). However, if we try to apply control theory to project organization, it may seem that all projects become inherently unstable by the time they reach the construction phase of implementation!

Pursuing the analogy, it could be argued that political aspirations – involving a conceptual or cultural project interface – represent positive feedback which will destabilize the project control loop. Such aspirations are a feature of many projects in newly industrializing countries, familiar to economists as Goodhart's Law: 'Any statistical regularity will break down if it is exploited by policy makers.'

INTERFACE MANAGEMENT IN OFFSHORE PROJECTS

Offshore projects involve the passage of very large amounts of information across organizational interfaces, both within the project organization and between it and organizations external to it. Careful control of these information flows is essential if management problems are to be avoided.

The most important interfaces are said to be to do with:

- the availability of funds for development
- the price of oil and/or gas
- taxation
- the size of reservoirs

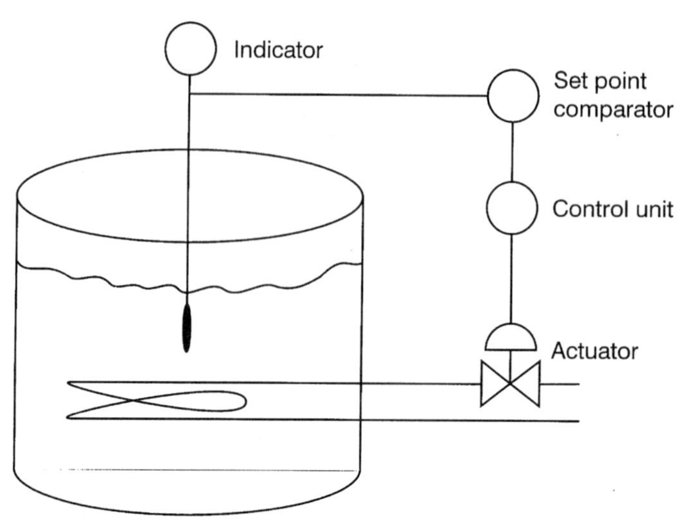

Fig. 6.3 The effectiveness of this control loop depends on the sensitivity of the sensor; the accuracy of the indicator; the ability of the comparator to provide an appropriate signal, and of the control unit to interpret the signal; the response of the actuator; and the amount of distortion (noise) in the transmission system. The same factors apply to project communications.

- the depth of water
- technology development
- Government attitude to development
- number and size of operating companies
- overall availability of resources
- experience, skill and mobility of engineering and management resources.

Most of these have analogues in onshore project development. Distinctions are of degree, rather than of kind. In both situations, conceptual design 'is the point at which massive prejudices tend to arise and dominate the discussion, and even obscure the correct choice, unless great care and diligence is exercised to maintain objectivity' (Gaisford, 1986). Remember that conceptual design is the stage when the project technology is chosen, a choice which determines some 80 per cent of the eventual costs committed to the project.

INTERFACES AND COMMUNICATION

Interface management is mainly a question of establishing rapid, reliable and accurate communications between the numerous organizations, groups, subgroups and individuals whose activities have to be coordinated to achieve the project's objectives. This requires firm leadership to establish appreciation at all levels of the importance of communication, appreciation that should be reinforced by effective control procedures.

Communication interfaces can be classified in a number of different ways. Basically, there are either *external* or *internal* interfaces in a project organization (*see* Fig. 6.4). The external interfaces may be between third parties, specific individuals, or subgroups within a project organization. For example, the interfaces between:

- project management and corporate management
- project engineer and design office management

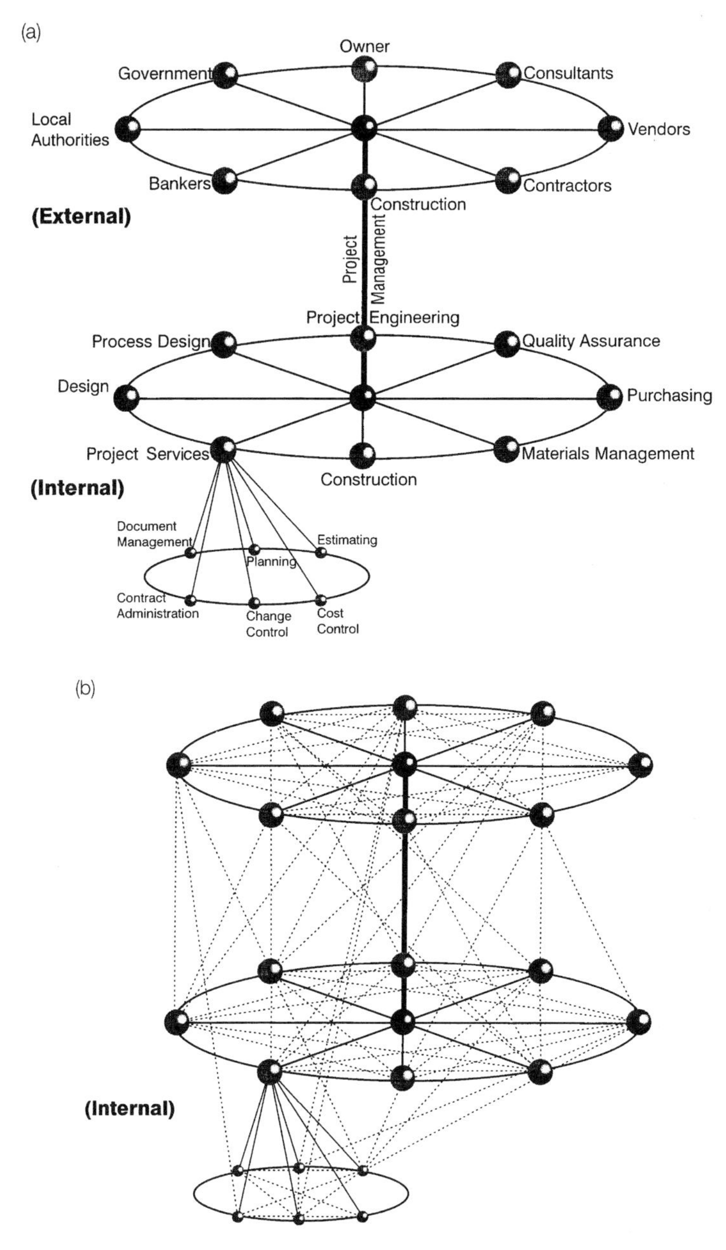

Fig. 6.4 (a) Responsibilities focus on project management. (b) Project management – communication interfaces.

- owner and contractor
- procurement management and vendors
- construction management and equipment or materials suppliers
- contractors and subcontractors, and so on . . .

Internal interfaces are those within the boundaries of the project team organization. Within a project team, obvious interfaces are those between:

- project services and project engineering
- cost control and planning functions
- project engineering and quality assurance, and so on . . .

In any project there will be several interfaces to be managed. The effective management of these interfaces is one of the most important tasks of the project manager – inefficient interface management is a prime cause of project failure.

INTERFACE GROWTH

In any project organization, all the organizational units have, in principle, the ability to communicate and exchange information between each other. The more units there are, the more communications interfaces come into play. Their number increases nearly exponentially with the number of players. For example, if n is the number of communicating units, then the number of interfaces I is:

$$I = (n-1)n/2$$

So, although it may seem obvious, it is worth emphasizing that the project organization can easily become more difficult to control as it gets larger. And the larger the number of contractors employed, the greater will be the division of responsibility, hence the more interface problems there are likely to be.

INTERFACE MANAGEMENT TOOLS

In addition to the proprietary software typically used to generate project progress information, such as schedule and expenditure tracking, resource levelling, etc., it is not difficult to write 'home-made' software tools to help in managing project interfaces. Some examples are:

- **'standard project report' forms,** which can be linked with a data base to generate automatic warning of follow-up or actions due. This facilitates communication within the project team. It also makes it easy to issue protocols of formal project meetings, showing decisions taken and action required, and 'deviation' reports showing all deviations from planned activities.
- **change forms,** standardized to speed up approval/rejection. These can also be linked to the project data base.
- **fax forms,** standardized for external communication. Linking to the project data base makes it easier to follow up necessary actions.
- **equipment lists and drawings** should be filed in such a way that all comments are retrievable. Final drawings should be microfilmed and stored safely.
- **meetings** are among the most important project management tools. They should always have an agenda (an open one, not a 'hidden' agenda), a chair person and someone else to take minutes (be aware that whoever takes the minutes is in a position to exercise undue influence on the outcome of the meeting!). The meeting must have a clearly stated objective. Copies of all relevant documentation should be provided in advance to everyone due to attend the meeting, so that everyone is fully informed by the time the meeting starts. And the meeting should have a definite duration: start and stop times must be respected by everyone who attends.

  Project meetings are typically held once a week, or more often at key periods. This implies that minutes of meetings must be produced and distributed *promptly*.

- **Constraints lists** are produced by anyone who is unable to do the job for which he is responsible. A constraints list sets out the reasons why he cannot do *his* job. The constraints must be resolved at the next project meeting. They must not be allowed to persist.

6.5 Organization and control

PROCEDURES

Organization, control and procedures go hand in hand. Procedures and documentation are among the basic control tools of the project team. There is a fine dividing line between providing an effective set of procedures and imposing a top-heavy bureaucracy which can kill incentive and seriously delay progress.

Procedures should be kept under frequent review and amended where deficiencies are discovered or circumstances change. The desirability of streamlining and simplifying documentation and procedures should always be kept in mind. Normally, it makes sense to use established procedures. However, they must be adapted as necessary to suit the particular circumstances of the project in hand. Attention to detail in the formulation of procedures and the design of documentation is very important. So is the need to ensure that the implementation of procedures and the management of documentation are carried out with meticulous attention to detail.

The administration arrangements on a large project are likely be complex. Slipshod and careless handling of data is a common cause of administrative delays. The use of computers for data processing and information handling makes accurate inputting of data of particular importance; otherwise spurious output can result, with the potential to cause serious difficulties.

The aim of control procedures should be to enable the project manager to 'close the books' every evening, i.e. to know day by day exactly what his commitments are.

RESPONSIBILITIES

The pıoject organization structure must be clearly defined. Usually this is done in the form of a series of organization charts. These should be widely published so that they are available to all members of the project organization and people who will interface with it.

The organization will change as the project moves through its various phases. It may also require amendment for other reasons. It is important that the organization charts are kept up to date, and that everyone concerned is kept informed about organizational changes.

Organization charts define lines of communication and responsibility. 'Terms of reference', setting out the specific responsibilities, authority and reporting relationships of all management and key personnel in the project team, should be published as an essential adjunct to the organizational structure. Much difficulty and confusion can be caused by failure to do this. Everyone concerned should be fully aware of their terms of reference.

LEADERSHIP

> This must be provided by the head of the project organization: the project manager or project sponsor, and likewise the managers, project engineers and others appointed with responsibility for controlling the separate project functions.

This, of course, is pure motherhood and apple pie. What – if anything – does it really mean? Well, we tried to say something useful about leadership (and management) in Section 1.2. And there is some more in Section 13.6, about leadership in certain person-to-person relationships.

Whoever is leading, at whatever level, will need to accommodate all these ideas. For while they will be mainly concerned with bringing together the efforts of many individual skills and disciplines, much more than mere 'coordination' is required of them. They carry direct responsibility for satisfactory completion of the project within the agreed schedule and budget. It is they who drive the whole project organization. Project objectives can be achieved only by firm and decisive leadership at all levels.

A great deal of project management time is spent in managing differences between people – 'conflict management'. Conflict and disagreement are unavoidable in project work because projects require the integration of work from many different individuals who may not usually work together and who may not normally report directly to the project manager. Yet some conflict is a good thing. It arises because people care about what they are doing (no-one fights about issues they don't care about), and it can ensure continued interest and energetic commitment. So you have to find ways to manage conflict constructively, rather than try to eliminate it. Usually, it is unwise to rely too much on logic and reasoning, because conflict situations are emotional as well as logical. Negotiation, focused on the common, agreed project goals, is probably the best approach. In any case, conflicts, like constraints, must not be allowed to persist.

But effective team leadership is not new. Neither is negative leadership. Petronius Arbiter, in AD66, said:

> We trained hard, but it seemed that every time we were beginning to form up into teams, we would be reorganized. I was to learn that later in life we tend to meet any new situation by reorganizing, and a wonderful method it can be for creating the illusion of progress while producing confusion, inefficiency and demoralization.

This, at any rate, we should try to keep out of our projects.

6.6 Summary

The overall organization of a project depends very much on the resources available to the project sponsor. There are relatively few types of overall organization, but the uniqueness of the project will demand a parallel uniqueness in the deployment of resources to plan and carry out the project. The main resource is people: their experience, training, work practices, and the ways in which their work as a team can be controlled.

The organization of the project team falls into two categories: the matrix organization in which the members of the project team retain their corporate identity and relationships; and the project task force in which, at least in principle, the project team becomes an autonomous group wholly dedicated to the project. The pros and cons of these alternatives tend to favour the task force for large projects, and the situation where the project team is working essentially in a 'contractor' role.

Because the project will pass through several phases, each having distinct objectives, the structure of the project team will adapt to suit the circumstances of the current phase. Throughout, however, the team needs a balanced membership in which personality will play as important a part as expertise or intellectual capability.

Motivating the team and organizing for project control fall to the project manager. A key issue is the management of communication interfaces between the project and the outside world, and within the project organization itself. The number of interfaces is liable to grow, when management becomes increasingly difficult. Procedures, largely designed to standardize communication, are useful as management tools, but must be employed with care to avoid bureaucracy. This, indeed, is a main function of leadership in organizing the control of the project.

7
Communication

7.1 Introduction

Projects are run – created, designed, built, operated – by and for people. People in an organization must communicate. An organization can hardly exist in the real world without people communicating in some way, but the criteria which characterize project work mean that they must communicate effectively and efficiently if the work is to be well directed and managed. One of the most frequent causes of project problems is communication failure.

7.2 Informal and formal communication

When we think about managing work, we tend to think mainly in terms of formal reporting lines. However, in any organization, informal communications networks also develop (Fig. 7.1). Individuals often communicate directly with people working in other branches of the organization in preference to formal channels, and often an individual emerges as a 'natural'

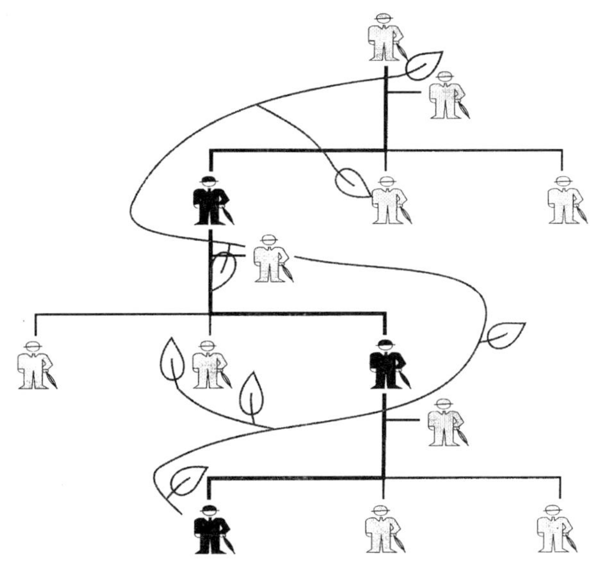

Fig. 7.1 Formal and informal communication.

communications centre. This could be because of a particular individual's position in the organization, or because he/she has particular personal qualities. The managing director's secretary is often the best source of information in the organization.

Informal lines of communication are important to the smooth running of the organization. But they must be controlled. In a project, in addition to cultivating informal communication, people must be aware of the need to adhere to formal project procedures, especially the need to ensure that essential decisions are confirmed in writing.

COMMUNICATIONS METHODS

There are a few methods of communication which apply to all types of projects (see below).

Spoken

This probably causes most problems. People are often unable to recall precisely what was said in a conversation or phone call. Messages which are important should be followed by at least a file note, or preferably a memo, recording the essentials of the discussion. All meetings must be followed promptly by the issue of minutes to all who attended and others concerned. The minutes should include a clear indication of who is responsible for follow-up action, and when it must be completed.

Written

There is always a conflict between the desire to limit the enormous quantity of correspondence which is a feature of large projects, and the requirement to ensure that essential communications are properly maintained. All members of the project team should be instructed to adopt the 'need to know' principle, so that information copies of written communications are limited to those who need the information. It requires experience and good judgement to get 'need to know' distribution right, but it is an essential of efficient project management.

The need for clarity and simplicity in written communications needs constant emphasis. Training in the art of clear, simple and accurate expression in writing is an important part of project staff training.

Graphical

Design and construction drawings are an essential basis for the whole enterprise. On a large project there will be many thousands of drawings to be issued, approved, possibly amended, reissued and filed for future reference. Consequently the procedures for the control of drawings form a vital section of overall project procedures. It is equally important to ensure that an adequate, properly staffed, administrative system is established, so that drawings can be managed efficiently.

Charts and graphs are also used to convey information. This particularly applies to planning and progress data. It is important that these data are presented in a clear and accurate way, and where necessary, supported by adequate written explanations so that ambiguity is avoided.

Numerical

Tables of data will also be used to convey information. Again, it is important that adequate explanatory notes are provided, so that readers get rapid and accurate understanding of the numerical information. Numerical data should be presented in a form which can be readily understood by management. For example, unedited computer print-outs are often difficult to comprehend without spending a lot of time and effort on them. Good editing assists understanding.

Electronic

Computers are in general use. They are very powerful and can deal with very large quantities of information. They must be used with intelligence and discretion. There is a natural tendency to believe implicitly in the output from computers. However, output should be regarded critically – it will have no more than the same degree of accuracy as the input. Be alert for cases, such as material quantities from computer-aided designs, where it may not be feasible to do a manual check of the figures if doubts as to accuracy arise.

Communication is useful only if it is understood. Keep the needs of your audience in mind when you prepare any kind of communication. Most of your audience, most of the time, will be under pressure and unable or unwilling to apply themselves to comprehending obscure or obtuse information. Try to make it easy for them to understand what you are telling them, especially if you need them to act on, or approve, your communication.

COMMUNICATION FACILITIES

Three main communication facilities are used by a project organization. Each has its own particular characteristics.

Reports, memoranda and letters

The need for clarity and accuracy has already been emphasized. Avoid the use of jargon, and only use acronyms where it is certain that they are understood by those concerned. The importance of minimizing the number of recipients has already been mentioned.

Memoranda and letters should each preferably be confined to a single subject, and it is helpful to use a running sequence of reference numbers for all written communications, retaining a copy of each memo or letter on a 'daily' file, in addition to the normal subject file copy. This is useful for tracing and retrieving papers quickly at a later date.

Telephone

The provision of adequate telephone facilities in all project offices and centres of operations is clearly essential. There have been many improvements recently in telephone technology, particularly in the introduction of portable instruments which can be used to make a project's communication system more effective.

The importance of recording important telephone conversations, particularly those involving decisions, has already been mentioned.

Electronic

It is important not to abbreviate telex, fax or e-mail messages to the point where misunderstandings can occur. Don't skimp at the expense of clarity. Otherwise, the same comments apply as to other written communications.

Both fax and e-mail are now widely used and can be of considerable assistance in conveying visual information rapidly between distant locations. Transmission 'noise' may distort the message. This is usually not much of a problem with words, unless the distortion is severe, but can be awkward when transmitting numerical data. E-mail facilitates very rapid communication. However, e-mail messages have considerable potential for going astray and getting lost. Insist on acknowledgement from the recipients of your e-mail messages.

Any communication calling for a reply should be promptly acknowledged. If you can't reply with all the information requested, say when a full reply will be made, and make sure that it is made then.

It is not expensive to provide adequate communication facilities, yet it is surprising how often corporate bean-counters will claim that the budget does not allow for enough equipment to be supplied to a project team. No project manager should acquiesce in such a lame excuse.

7.3 Documentation

GENERAL

All projects generate paper. Project *reports* make up a large part of the paper mountain. Some become part of the project design basis, some may come to form part of contract documentation, and some are simply information dealing with, for example, progress of work (or lack of it). Some reports contain proposals for new or revised work, and may have to be submitted for review and approval by higher authority – a necessary but potentially time-wasting process, dealt with in Section 7.4, 'Reviews and approvals', below.

Project *schedules* will be required, and it is important to make sure that the level of detail given on a schedule is appropriate to the level of management for which it is intended. A master schedule with, say, not more than a hundred activities may be suitable for overall control. It would be supported by subnetworks covering different parts of the project in detail.

Control, particularly during the design stage of a project, is made easier by frequent issue of priority lists ('to do', or 'punch' lists). Issued, for example, weekly, they list the essential decisions and actions required by the programme during the coming week. Firm control of the production, approval and issue of *drawings* is essential, not least to control design changes. Uncontrolled design changes are one of the most frequent causes of project delay and extra costs. When a revised drawing is issued, make sure that everyone affected by it knows about the revision, otherwise somebody will continue to work on an out-of-date drawing.

There must be a proper system for the production, approval, recording and filing of all *calculations* associated with the project designs. Procedures for this should form part of the overall project procedures. This is particularly important where parts of the installation have to be certified by an independent inspection agency.

Specifications should be comprehensive, clear and precise, and as far as possible open to only one interpretation. Frequently, specifications will refer to national or international standards. Check that such references are fully relevant to the particular specification concerned. As with other documentation, it is important that the production, approval, change approval, distribution and filing of specifications form part of the project documentation control procedure.

Manuals have to be provided for the plant and equipment to be installed in a new project. They will come in the form of vendor's data, equipment manuals, spare parts lists, operating instructions and similar documentation. Requirements for these must be indicated in the specifications.

PROJECT REPORTS

Project reports are the main formal means by which project progress is communicated and recorded. As records, they are vital parts of the project's audit trail, so they must be prepared with all possible care. Project reports should:

- state clearly the current status of the project;
- compare actual achievements with the planned target achievements;
- draw attention to critical issues;
- identify problems and propose solutions;
- promote effective management and control.

Effective management and control needs *timely* action, so project reports must be produced regularly and promptly (e.g. monthly reports should be issued no later than 10 days after the 'current status' date). It is usually better to have roughly accurate information quickly than to have precisely accurate information late. Project reports should preferably be based on 'exception' reporting, i.e. highlighting variances from the plan. This helps the reports to be (1) concise, and (2) forward looking.

Much project reporting is concerned with schedule and budget control. Project reports should draw attention to overruns as well as savings, together with proposals, or action already taken, to remedy the former.

Reporting on the overall progress of the project is done on the basis of:

- earned value of work done versus work planned;
- milestones achieved and planned, effects on the schedule and budget;
- resources required and available (e.g. current month; next month; next year);
- performance indicators such as measured work completed and labour productivity.

Project reports must be adapted to the needs of the recipient. Reports for corporate management will be different from reports for technical functions and those for joint venture partners. But in any case, a report should be easy to understand. Graphical information is often better than words or numbers in this respect. Project reports should be written with a view to the results (decisions, actions, authorizations etc.) that they are intended to achieve. Using standard layouts promotes this.

As the project passes through its various phases, some types of report become redundant. Check the need for particular kinds of report frequently and continue only those which are relevant.

Finally, when the project is complete, the project manager must write a *debrief report* covering at least:

- performance with respect to the planned/authorized budget;
- performance with respect to the planned/authorized schedule;
- use of resources;
- aspects of the management and execution of project work, particularly what he did and would do again; what, with the benefit of hindsight, he would do differently; or what could have been done better.

Material for the debrief report should be assembled from progress reports and project records written during execution, and should not be put off until the end of the project when recollection may be imprecise or coloured by past events.

REPORTING AND INFORMATION SYSTEMS FOR PROJECT IMPLEMENTATION

Report types, formats, frequency and required contents are specified in the project implementation plan. Implementation reports are intended to:

- identify clearly the current status of project work;
- compare performance achieved ('actuals') with targets ('plan');
- focus on critical areas;
- indicate proposed or accomplished solutions to the problems identified;
- be prompt so that effective corrective action can be taken in good time.

The main implementation report types are:

- progress reports
- cost reports
- earned value reports
- project debrief reports (an especially important type of report, for they provide the means of learning how to improve the management of future projects).

While report formats are usually standardized to conform with corporate procedures, all reports – standardized or not – must be carefully planned. Have in mind the following: tailoring the report to the needs of the reader(s); making it readable and easy to understand; highlighting problems *and* proposed solutions; the response/action to which the report is expected to lead; maintaining a permanent reference point against which to measure long-term performance; replacing words, so far as possible, by tabulated data or graphs.

Information systems

Large projects generate very large quantities of information. If this information is to be used effectively to manage project work, it must be organized systematically and structured to provide suitable information for the various levels of engineering, project and corporate management. The main project information requirements are for:

- cost estimating and cost control
- scheduling and schedule control
- contract management
- purchasing
- quality assurance
- design management
- documentation management
- compiling the equipment register.

Computer based project management information systems (PMIS) are widely available. They can be tailored to suit particular needs, and can be networked to contractor or site offices. Configuring the PMIS requires detailed definition of the requirements and processes in which information will be used. However, since the major contractors operate comprehensive PMIS, in projects where a turnkey or a managing contractor is retained, the first option to consider is the use of the contractor's PMIS.

CONTRACT DOCUMENTATION

The clarity and freedom from ambiguity which arise from careful drafting are essential features of all contract documentation. Without them, the project will become an expensive pasture for lawyers to graze in.

Invitations to tender (ITBs) must contain a complete specification of what the client (sponsor or owner) requires the contractor to do. *Tenders* should contain a complete description of what the contractor is offering, and for what financial consideration. Generally, before a bid is accepted by a client, there will be detailed discussion of its contents, followed by an amended bid. This process is intended to ensure that the client is satisfied that the services being offered will meet his requirements.

The *contract* is the document which legally binds client and contractor together, and which governs their working relationship. The contract will include at least the project scope of work, technical specifications, construction drawings, project schedules, conditions and terms of payment. Other associated documentation, such as bills of quantities, schedules of rates and formal control procedures, may also be included.

It is not always easy to ensure that all the parts of the contract documentation are mutually compatible. When different parts of the contract are prepared by different parts of the organization you, as project manager, need to pay special attention, *before the contract is finally agreed*, to ensuring that what is said in one part is not contradicted in another. Control of the important interface between the client and the contractor is based on this documentation. It is therefore critical to the smooth running of the project.

7.4 Project procedures

GENERAL

Procedures provide the framework for control of the project. They govern the ways in which information flows between the various parts of the project organization. When they work, they are the main interface control mechanism. When they do not, they are a major cause of delay.

Project procedures should never be considered sacred, but they should be kept under review. They should be amended to suit changing circumstances, and procedures that do not work well should in any case be changed. A project is dynamic: the situation changes daily. Procedures should adapted accordingly.

Project procedures require intelligent interpretation. They should not be allowed to become a bottleneck which inhibits initiative and stifles action. We have to aim to maintain a balance between the need for effective control and excessive bureaucracy. They should be streamlined as far as possible. If it is proposed to incorporate old established procedures in a procedure for a new project, take care that they are completely relevant to the new project situation. Large organizations which are continually carrying out projects will have sets of standard 'boiler-plate' procedures which can be rapidly assembled for a specific project. (This particularly applies to the international engineering contractors and to some of the large oil companies.) Assembling the 'boiler-plate' must be done with careful thought and sensitivity to the needs of the project.

A special feature of some project communication procedures is *confidentiality*. Much project information, both of client and contractor origin, is considered to be confidential by one party or another. For procedural purposes, confidential information should be clearly marked, and special provisions made for its transmission and storage. Two points worth keeping in mind: (1) if every document is marked CONFIDENTIAL or SECRET, then no document will actually be considered so; (2) telephone, telex and fax transmissions are seldom confidential.

HANDOVER

Handover marks the transfer of responsibility for the project from the contractor (or the department acting as contractor in the case of a wholly in-house project) to the client (owner or sponsor, or the in-house department which will operate the project facilities as a business). Procedures consequently count for much in the handover process. Handover begins with pre-commissioning, and ends when the final acceptance certificate has been signed and the project has been *closed out* as detailed below:

- **Pre-commissioning** covers checking the integrity of equipment and systems, at vendors' works and at the project site, before and after installation. In offshore projects, maximize onshore testing as much as possible.
- **Hook-up** covers the connecting up of systems and equipment after installation so that the installed facility is operable. The project manager should make sure that hook-up plans and checklists are prepared for every system.
- **Commissioning** covers work at the project site to bring the installed facility into a state of 'ready-for-use'. Typically, this is formalized in a detailed agreement between the project manager and the client, which is the basis for the client to take over the project site. Commissioning remains the responsibility of the project manager.
- **Start-up** includes the provision by the project manager of the resources to carry out trouble-shooting, testing and performance demonstration. Start-up procedures are prepared by the project manager and agreed with the client. The project site is now the responsibility of the client.
- **Handover** includes transfer of all operating and test documentation to the client. Handover is sometimes phased: installation handover immediately before start-up; final handover when start-up has been completed, the hand-over report accepted, and operational and test documentation accepted.

The details of the handover procedure must be agreed beforehand by the client and the project manager. If contractors are involved, the details of commissioning and post-commissioning arrangements must be negotiated and defined before contract award.

Handover is complete only when full project documentation, including operating and maintenance manuals, are available and 'as-built' drawings are complete. However, the project manager has not finished his task until the project is *closed out*. This requires him to check against the budget:

- un-invoiced work done or materials received
- invoiced but as yet unpaid items
- claims and surplus materials.

The project manager then prepares a final cost report and the project debrief report. The project is complete only when:

- handover has been completed;
- all project documentation is complete;
- administrative documentation has been scrapped or archived;
- surplus materials have been scrapped or otherwise disposed of, and accounted for;
- initial spare parts have been received into store;
- standards which deviated from the basis for design have been updated;
- the budget has been closed;
- the project debrief report has been written.

REVIEWS AND APPROVALS

Reviews typically take place at the conclusion of each planning stage and throughout the implementation phase, to coordinate project activities and to check that work is going according to plan.

Process development reviews consider the latest technological developments likely to affect the project. They are carried out during project identification and feasibility studies.

Technical reviews assess the technical design and specification at various stages, so as to ensure the quality and technical integrity of the project. They include reviews of the basis for design, project and technical specification reviews, and hazard and operability reviews. They are carried out at various stages during the definition and implementation phases.

Milestone reviews assess progress at critical stages of the project. Their purpose is to check that the project is still viable before proceeding to the next stage. They are carried out by considering:

- *past activities*: was the plan followed/changed; what were the consequences of change?
- *present status*: is the project currently viable; what are the risks; what uncertainties remain?
- *future activities*: what is the plan for the next stage; will the required resources be available at the specified time; what steps must be taken to ensure this, or suitable alternatives?

Milestones are often used as the triggers for payments to contractors. They should therefore be defined with considerable care.

Management reviews provide an overview survey of the status and viability of the project at critical stages. They are similar to milestone reviews, but concentrate on particular features involving uncertainty or risk. They are most important at stage completions during the definition phase and, because of their role in determining that the project is properly planned for progressing to the next stage (and thence to implementation), they are best carried out by an independent group of experienced people.

Project reviews take time and incur costs. Reviews must be allowed for in planning the project schedule and budget. Reviews provide a means to check, consider and decide before going on with project work. Very often, a review is a necessary precondition for requesting approval for funds to continue to the next stage. Getting approval also takes time – which must be allowed for in planning the project – and is made easier by standardizing request-for-approval procedures and documentation.

A typical procedure is as follows:

- approval is to be requested only for well-defined activities (a review beforehand would help ensure this);
- one person is responsible for obtaining the approval ('single-point responsibility');
- approval for a particular activity or set of activities may be requested only once.

Typical guidelines for approval documentation are:

- use standard forms;
- design the forms to present information clearly and concisely;
- present the information so that whoever has to decide on the approval has all the information he needs, unambiguous and easily understood, so the decision can be made promptly.

Special considerations apply to *handover* and to *statutory* obligations. Handover documentation records what is in effect the client's (or operating department's) approval and acceptance of the project facilities, and, since handover usually has special contract significance, the nature of the documentation will usually be specified in the project contract.

Statutory obligations may require approvals from various authorities under, for example, planning and development legislation; environmental regulations; health, safety, hazard and emergency management regulations; works and equipment certification, etc. Details of requirements depend on where the project is located. The project manager needs to get guidance about local requirements as early as possible in the development of the project so that suitable procedures for documentation can be put in place in good time. (Approvals are also discussed in Section 15.6.)

7.5 Causes of communication problems

There are many causes of communication problems on a project. Some of the more common are discussed below.

LOCATION

The location of the project offices (of the client company, and of the contractors and other participants) is important for maintaining effective communications. Badly sited or unsuitable offices can have a deleterious effect on project communications. Even allowing for modern communications technology, physical proximity can be a significant benefit for the efficiency of communications.

The client's office

Where the project manager is located will depend on the circumstances of the project, and may also be influenced by the policy of the client organization. The availability of suitable accommodation will also be a factor, but should not be allowed to influence the choice of location to the detriment of the management of the project.

Corporate management may well wish to keep the project manager near at hand. He, however, will probably want to be near the contractors. When a project is being undertaken in a country other than that in which the client's headquarters are based, the project manager will often be located in that country.

Design work is often undertaken in a different country from that in which the new installations are to be built. In this event, the project manager may change his location when the project moves into the construction phase.

Access to good transport facilities is important. Proximity to an airport and surface transport networks facilitates rapid travel between different project locations.

The contractor's office

Typically, design contractors have offices at main centres of population. However, in some cases, the design team might be located at, or near, the construction site. This is usual where extensions to existing plant are involved.

There may be political pressure to locate the whole project organization in the country in which the new installations are to be built. The closer together the various elements of the project are located, the easier will it be to maintain good communications. For example, in a recent project, the client leased a disused factory and located the whole of his and the contractor's teams under one roof. The cost of this was high, but the project was reported as being very successful in terms of early completion and reduced cost. The client no doubt considered that the extra expense of housing the whole team under one roof was justified.

OFFICE ACCOMMODATION AND FACILITIES

Project offices

Some of the factors which need to be considered are as follows:

- Are the project offices adequate to accommodate the organization throughout the life of the project? Project teams invariably grow, often larger than was expected.
- Are the office arrangements flexible enough to permit the layout to be altered as the organization changes?
- Careful allocation of offices and space is necessary, particularly bearing in mind those who need to communicate together most frequently.
- Try to arrange the offices to encourage direct personal contact. 'Open plan' offices, while seldom popular with staff, are good in this respect, and can be made more congenial by intelligent use of partitions and potted plants.

Facilities

Don't skimp on providing adequate facilities in all areas. The following are important in this respect:

- **Secretarial services** – ensure that they are adequate to provide proper service to the project team. The same applies to *telephones, telex and facsimile services.* Radio may be required in certain cases. A courier service to ensure prompt delivery of mail may be justified.
- An adequate administrative organization is required to deal effectively with documentation control. Trying to save money by using low-grade clerical staff for this job often leads to expensive disasters.
- The early appointment of an efficient office manager is essential if the administrative organization is to function smoothly and good communication control is to be maintained.

WEAK PROJECT LEADERSHIP

We have already emphasized the importance of effective leadership. Good project procedures supported by a good administrative system will not compensate for weak project leadership. However, there are actions which can reinforce project leadership and make it more effective.

Publishing the overall project programme and ensuring that the project objectives are clearly understood at all levels will help to strengthen team spirit and improve the efficiency of the project team.

Communication skills should be promoted by management. The project manager should ensure that everyone is aware of the need to give attention to:

- correspondence, especially prompt acknowledgement and reply;
- reports;
- management of meetings;
- interface management.

If key personnel in the project team are not performing adequately, then consider changing them. Alternatively, lack of adequate training may be the cause of poor performance. If this is the case, arrange suitable training (quickly: you don't have much time once the project is running!).

ORGANIZATION AND DISCIPLINE

Poor project discipline is often associated with lack of organization. Often the root cause is weak project management. Possibly, the management team needs strengthening, or perhaps the project manager is overloaded. It may be necessary to change key personnel or to make changes to the organizational structure.

Sound quality assurance procedures, properly implemented, help to ensure that project discipline is maintained, as well as performing their control function.

As mentioned earlier, it is important that the project organization is published. It is also important that this information is kept up-to-date. There will frequently be changes to the organization as the project develops. These should be published as they occur, so that everyone is aware of the changes. Often project organization charts are out of date and do not accurately reflect the way the organization has developed. This leads only to confusion and inefficiency.

Key posts in the project organization should have written terms of reference, with clear definition of responsibilities and reporting relationships. The terms of reference should also be published and circulated to all members of the project team. These documents should be kept brief and succinct. In the case of management appointments, it is an advantage if people write their own terms of reference, and then agree them with their immediate superior. People are then more likely to be fully committed to their assignment.

Ultimately, it is for the project leader and the management team to provide guidance and set an example in all aspects affecting organization and discipline.

PERSONALITY AND INTERDISCIPLINE CONFLICT

Personality clashes occur from time to time in all organizations. People often behave in temperamental, emotional or irrational ways. In designing and staffing a project organization, it is especially important to take account of individual personal characteristics (*see* Fig. 7.2). Try not to 'fit square pegs into round holes'. The project organization must be balanced. Flexibility, a willingness to adapt the organization to the people available to fill it, is important, but should not be allowed to degenerate into limpness or vacillation.

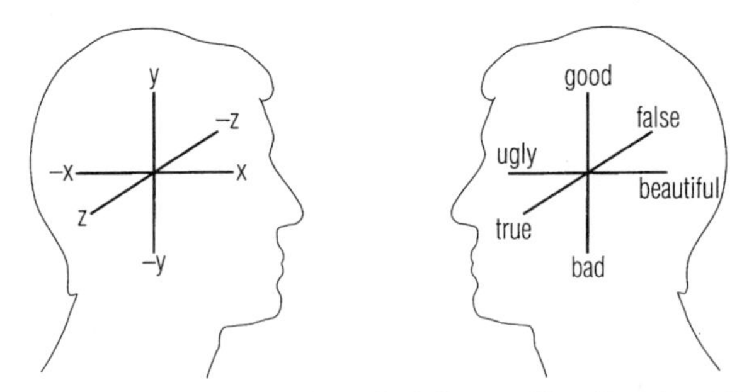

Fig. 7.2 Communication: personality and interdiscipline conflict. (Based on Themerson, 1974.)

While it is unwise to put people whose personalities clash to work closely together, it is sometimes difficult to avoid interdiscipline conflicts. For example, there seems to be a natural antipathy between site construction staff and the project engineers and designers. The latter are responsible for ensuring that drawings and materials are available when required at the site. Construction managers are, by calling and necessity, people with strong personalities, so they sometimes appear to resent the overriding authority of the project manager.

It is helpful if project engineers have some site experience. Similarly, construction staff should ideally have had experience in a design office. Conflicts between design and construction are then less likely to occur. Teach-ins, seminars and information meetings are a way of ensuring that staff can appreciate each other's responsibilities and problems. It promotes team spirit, and helps to minimize interdisciplinary problems. Nonetheless, firm leadership remains essential.

If problems arise which are due to personalities or other 'people' causes, then prompt action is essential. Otherwise the smooth and efficient running of the project can quickly be jeopardized.

INEXPERIENCED PERSONNEL

Inexperienced personnel can be accommodated in the project organization, but should be carefully placed where they can gain experience in jobs within their capability.

Time should be found for training. Often we find that people with great potential rarely get any formal training because they are always in demand, and – their line superiors say – cannot be released. However, practical experience is also vitally important. Theoretical knowledge is valuable, but 'hands on' practice is essential for people assigned to the project team. This means that senior project people must find time to act as mentors to their subordinates. It is always difficult to squeeze more into the standard 25-hour project day, but mentoring is time very well spent, for 'It is only when you undertake to use what you have learned that you discover whether you can transform knowledge into performance' (Nero Wolfe: Stout, 1995).

INADEQUATE DOCUMENT MANAGEMENT

Project information must flow quickly to its destination. Accuracy in all types of communication is crucial. Clarity, brevity and accuracy should be encouraged – indeed, insisted on – at all levels. The sensible use of priority codings on correspondence is also important, but if every document is addressed as 'urgent' then the coding becomes valueless.

An efficient storage and retrieval system is also vital. The project team should include enough, and suitable, people to handle the large quantities of documentation that are inevitably generated in a project. Experts can be used to devise efficient systems, and modern office automation can be used to reduce the paper burden and speed up communications. But both experts and systems will be negated if documentation management is not actually done by well trained, well motivated and reliable people.

LACK OF COMMITMENT

Lack of commitment in the project team will adversely affect communications. In fact, a breakdown in communications may be one of the early indicators that the morale of the project team is falling.

Maintaining morale when a project is in difficulties is a searching test of leadership. Giving praise where it is due is important in encouraging people to maintain their commitment to project objectives. Ensuring that everyone is aware of the project objectives, the project programme and the current situation will assist in promoting the team spirit. All members of the team should be aware of their contribution to achieving the project objectives.

So far as it is possible, the project manager should ensure that his staff are adequately rewarded, having regard to levels of responsibility, work load and performance. Congenial working conditions are also important. Project accommodation is often of a temporary nature and, unless positive steps are taken to improve it, leaves much to be desired in terms of a reasonable working environment.

It is crucial that approval times and deadlines are respected by all concerned, both by clients and contractors. But people should be set realistic work targets, particularly regarding completion times for tasks. Continually setting unrealistic targets eventually destroys commitment and lowers morale.

An occasional internal project meeting, to which all project staff are invited for a briefing on the up-to-date project position, is useful in maintaining morale. Those invited should include not only the management, but also secretaries, administrative assistants and filing clerks, indeed everyone concerned with the running of the project.

Social contacts can be helpful in promoting the team spirit. For example, it is good to arrange occasional social events, to which wives or partners are invited. Project people are frequently required to work long hours and to make unplanned absences from home. Wives and partners should understand and appreciate why. They are then more likely to support the project team member's commitment to his job.

While this chapter has been concerned mainly with communication, the importance of leadership and management of the project organization has been emphasized. It is the most important ingredient of a successful project.

7.6 Summary

Communication is what makes organizations work. While we are used to thinking of communication in the formal structure of an organizational hierarchy, informal communication – which bypasses the hierarchical structure – is important and useful if controlled.

Whatever the method of communication, it will be useful only if it is understood. We need to adjust our communications to the needs of the receiver, especially when the aim of the communication is to elicit a required action, as well as to inform. From the project point of view, we have to ensure that adequate communication facilities are put in place and properly used.

Sooner or later, project communications come to exist as documentation. Documentation management therefore becomes a key area of project management. Contract documentation and project procedures are particularly important.

Communication problems cause project problems. Some of the more important communication problems relate to location, office accommodation and facilities, organization and discipline, the involvement of inexperienced people and, especially, weaknesses in project management. Correcting the latter is the key to solving the other problems.

8
Estimating Project Costs

8.1 Introduction

Costs must be estimated for any project. Project leaders and project managers themselves do not usually make detailed cost estimates: this is the job of specialists. But because cost estimates are so important in any project, all project people need to develop an almost instinctive 'feel' for the soundness of the estimates they work with. In this chapter, we consider how costs are broken down for analysis; the principles of value engineering and cost engineering; and we look at the way in which cost estimates are made and used.

COST BREAKDOWNS

Cost estimates are broken down in various ways. This makes it easier to make the estimates and to analyse them when made.

CAPEX and OPEX

CAPEX (i.e. 'capital expenditures') includes the charges on the capital investment required to realize the project. OPEX (i.e. 'operating expenditures') is the total cost of running and maintaining the project facilities. Depending on the jurisdiction to which the project is subject, CAPEX and OPEX may be treated differently for taxation purposes. The difference is important in profitability analysis.

On-sites and off-sites

On-sites typically include all the facilities for converting raw materials into products. Off-sites include facilities which support the functioning of the on-sites, such as utilities generation and distribution, storage and infrastructural facilities.

Off-sites tend to be neglected in cost estimating, perhaps because process engineers (who are responsible for the original project concepts) find them less technically fascinating than the process technologies which lie at the heart of the on-sites.

Logistics

This refers to the systems which move raw materials etc. to the project, and which move products, by-products etc. from it to the customers. The costs associated with logistics tend to be not so much neglected as forgotten, although careful design of logistic systems can contribute significantly to both CAPEX and OPEX savings.

Note that a sizeable part of a project's *working capital* will be associated with logistics. Working capital is the money needed to sustain production until revenues from product sales are received. Estimation of the cost of working capital is notoriously haphazard.

Variable and fixed costs

Variable costs are those which change roughly in relation to business activity. An example is the cost of raw materials per operating year (or other operating period). If you produce less (or more) product, you consume less (or more) raw material, provided yield is constant. The 'shape' of the relationship depends on the nature of the production process. *Fixed* costs are those which do not change in relation to business activity (*see* Fig. 8.1).

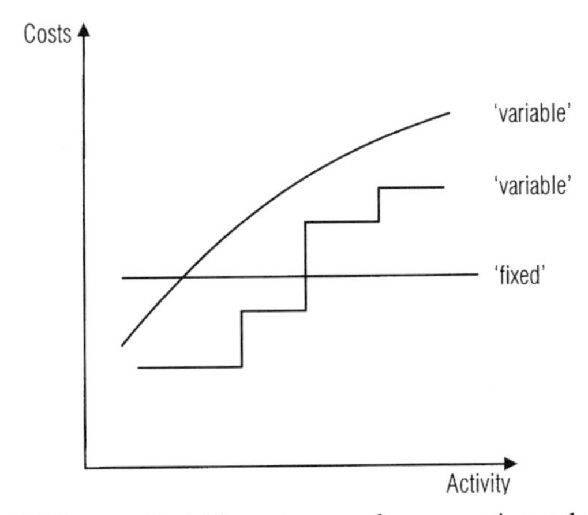

Fig. 8.1 'Fixed' and 'variable' costs. Variable costs may change continuously or stepwise in relation to activity.

The classification of costs as variable or fixed is determined by corporate practice. A fairly typical, but by no means prescriptive, classification is as follows:

- variable costs
  - raw materials
  - utilities and services
  - consumables
  - maintenance materials
  - spares and spare parts
  - charges on working capital
- fixed costs
  - operating and maintenance labour including supervision
  - general management overheads
  - payroll and plant overheads
  - royalties and license fees, unless based on output (when they will be *variable*)
  - insurances
  - depreciation
  - taxation
  - charges on fixed capital.

VALUE ENGINEERING

The main objective of value engineering is to find the minimum investment cost necessary to achieve a specified technical performance with specified reliability. Value engineering aims to find out and get rid of anything in design which is unnecessary, while maintaining or improving the performance quality of the process or plant under consideration.

The detail of technical analysis demanded by value engineering is considerable and expensive, and the resulting benefits are not always predictable. For these reasons, value engineering has sometimes found less favour in the process industries than in, for example, automotive manufacture.

Traditionally, the process industries have tried to reduce investment costs by:

- applying new technology, usually involving fundamental changes in design;
- employing cheaper materials of construction or reducing engineering standards.

While the former is acceptable, the latter often is not. But the picture is far from being black and white. The initiative to reduce costs is typically divided between the engineering contractors, responsible for designing and building a plant, and the owner organization responsible for operating it. We may note here that, under a typical 'lump sum' (i.e. 'fixed' price) contract, the contractor will have no interest in reducing the owner's investment by applying value engineering unless the owner – whose profitability could be significantly improved by value engineering – takes the appropriate steps. These include:

- developing in-house process engineering design and cost estimating expertise;
- improving the functional definition of project scope and design options;
- establishing clear design concepts, eliminating over-design and controlling design changes, including random efforts at cost cutting (which usually result in eventual cost overruns);
- establishing definite quality standards and the means of ensuring that they are met.

These steps, as a whole, are simply the application of project management principles to the early design stages of the project. This is 'value management', applied where it will have most effect, for most of the total project investment is committed by decisions taken at the conceptual stage when, regrettably, corporations are most often inclined to penny-pinch on resources.

COST ENGINEERING

An important instrument for project planning and control is the relationship between expenditure and the period of time through which the expenditure is committed, i.e. project cost in relation to project time (Fig. 8.2). The design and construction of these cost/time relationships, of accuracy commensurate with their purpose, is termed *cost engineering*.

The steps in cost engineering are as follows:

1. Define the nature and the scope of the project.
2. Break the project down into 'building blocks', to the level of detail appropriate for the type of estimate required. The work breakdown structure (WBS) is a convenient basis for this.
3. Break down each 'building block' further to identify the respective project deliverables. (The WBS makes it easy to allocate responsibility for each deliverable to named individuals.)
4. Estimate the cost (i.e. the hardware plus the work that has to be done on it to turn the hardware into a fit-for-purpose project deliverable) of each deliverable. This is the 'raw estimate' cost of the deliverable.

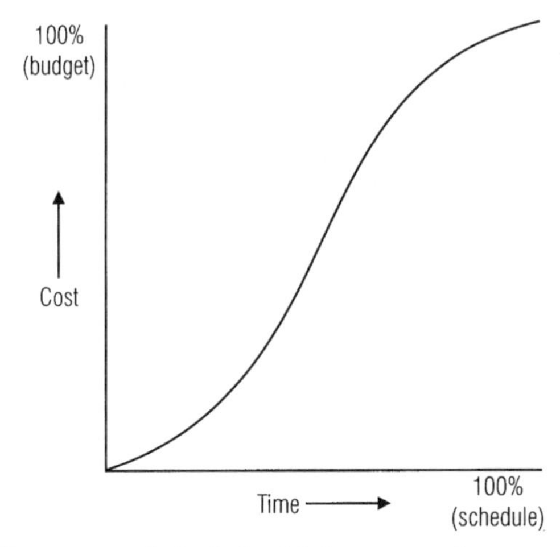

Fig. 8.2 The 'S' curve: a typical relationship between project cost and project time.

5. Add appropriate allowances and contingencies to the raw estimates for each item in the breakdown.
6. Develop an expenditure profile by phasing the components of the estimate in accordance with the project schedule.
7. Summarize and record the complete estimate from scope definition, through addition of allowances and contingencies, to the final project cost estimate.

These steps should be carried out for all types of estimate. However, the extent of detail of the project definition and the breakdown will be determined by the purpose of the estimate and the accuracy required. The required accuracy of an estimate is determined by the purpose for which it is intended. (Using an estimate prepared for a purpose such as project screening for another, higher level, purpose such as project budgeting, is one of the more heinous crimes to which project champions are prone!)

8.2 Purpose of estimates

Estimates are required for:

- project economic appraisal;
- studies on the optimization of design, implementation and operation;
- assessment of the probable impact of risks, changes and alternative courses of action;
- establishing budgets for project funding;
- providing the basis of project cost control during implementation;
- providing the basis of final cost projections.

For each purpose, a different level of estimate accuracy is appropriate. This leads to the classification of estimates by 'type'.

TYPES OF ESTIMATE

In this chapter, estimates are classified in terms of their purpose and accuracy, eg:

- type I: screening estimate, ± approx. 40 per cent

- type II: study estimate, ± approx. 25 per cent
- type III: budget estimate, ± approx. 15 per cent
- type IV: control estimate, ± approx. 10 per cent or better.

The accuracy of an estimate depends mainly on the amount of engineering effort permitted by the project scope definition. It is also influenced by the information used in steps 4 and 5 above.

Several other classifications exist, although apart from the terms used they are conceptually broadly similar. For example, a classification by the Association of Cost Engineers (ACE) is:

- **Class I – 'Execution'.** Design detail as used in project execution or definitive bids by contractors. Estimate based on material take-offs (MTOs), detailed labour and material breakdown, with correspondingly low contingency.
- **Class II – 'Design'.** Design detail as for preliminary estimates and budget costs, capital authorizations and project evaluations. Estimate based on reasonably thorough studies and calculations, with a medium level of contingency.
- **Class III – 'Outline definition'.** Design detail as for order-of-magnitude studies and feasibility study level of evaluation. The estimate is typically based on historical cost relationships and service data. Contingency will be correspondingly high.
- **Class IV – 'Conceptual definition'.** Screening estimate for the rapid assessment of commercial opportunities. Estimate based on conceptual engineering data, usually shown *without* contingency.

ACCURACY OF ESTIMATES

Estimates are forecasts of an uncertain future. Uncertainty is catered for by adding amounts (i.e. allowances, contingency) to the estimated cost of project items and the associated functions. In effect, this reduces the probability that the actual project cost will overrun the estimate.

The expected accuracy of an estimate can be expressed as a bandwidth in relation to the estimated cost which the project is equally likely to overrun as to underrun (hence referred to as a 50/50 estimate), with a probability of less than x per cent of overrun and underrun. For example, if we set $x = 10$ per cent, then a 50/50 estimate of 100 units, with an accuracy bandwidth of ±25 per cent, has (according to this practice) a probability of less than 10 per cent that the actual cost would exceed 125 units, and a probability of less than 10 per cent that the actual cost would be less than 75 units (assuming that the probability distribution is not skewed) (*see* Fig. 8.3). Another way of looking at it is to say, we have [100 – (10+10)] = 80 per cent confidence that the actual project cost will fall between 75 and 125 units.

8.3 Allowances and contingency

The amounts added to estimates to cater for uncertainty can be classified as 'allowances' or 'contingency'. They may be added to individual items in the project breakdown or to the project as a whole.

Because the effect of these additions is to increase the probability that the final (actual) project cost will not exceed our estimate, our confidence in the estimate tends to increase as allowances and contingency are increased. But unnecessary additions for uncertainty may lead to potentially good project opportunities being missed. On the other hand, additions which are too small lead to overoptimistic expectations. Consequently, great care is needed in applying these add-ons in project estimates.

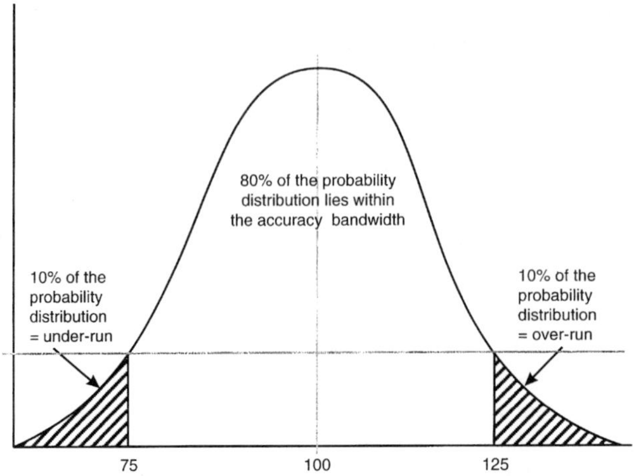

Fig. 8.3 Accuracy and confidence in estimates.

There are three kinds of such add-ons:

- **Activity allowances,** added at the discretion of the estimator to account for 'known unknowns' such as cut-and-waste, lost time due to weather conditions, re-engineering, weight growth in offshore structures, etc., which are known by experience to be very likely to occur. The estimator's raw estimate plus activity allowances produces the 'base estimate'.
- **Contingency,** added to the base estimate to account for 'unknown unknowns' such as variations in the cost of design and implementation which are within the agreed scope of the project, but which cannot be quantified when the estimate is made. Examples include the impact of market forces on the availability of resources and the effect of currency exchange rate fluctuations. Contingency is typically added at the discretion of the project manager, taking account of the estimator's experience. The resulting estimate is the 'most likely' estimate, i.e. the 50/50 estimate. By definition, the project is as likely to overrun this estimate within its accuracy bandwidth as it is to underrun. The 50/50 estimate is used as the basis for all project management purposes *except* for 'minimum risk' situations (see below).
- **Overrun allowance,** added to the 50/50 estimate to allow for the risk of overruns. Typically, the overrun allowance is an amount sufficient to transform the 50/50 estimate into a 90/10 estimate (i.e. an estimate with only 10 per cent probability of overrun). The 90/10 estimate is used as a 'minimum risk' parameter in setting corporate budget levels and in project economic sensitivity analysis. The overrun allowance is not normally available to the project manager except with the approval of higher corporate authority.

ESTIMATING THE AMOUNT OF ALLOWANCES AND CONTINGENCY

Recommended practice is first to estimate the allowance or contingency for each item in the project breakdown (step no. 5 in 'cost engineering' above). Different contingencies, as percentage on-costs, may be assigned to different items. If there are many similar items (e.g. repetitive activities) in the project, it may be acceptable for the contingency on several of them to be less than would be the case if there were only one such item, since the overall uncertainty is likely to be reduced by repetition.

The individual item contingency percentages are converted to absolute costs, summed and then divided by the base estimate. This produces the overall contingency for the total cost estimate. For example, overall contingency levels for the estimate types and accuracy bandwidths are typically as shown in Table 8.1.

Table 8.1 Overall contingency levels and accuracy bandwidths for the different types of estimate

Estimate type	Accuracy (±)	Overall contingency
Type I: screening	40%	20%
Type II: study	25%	15%
Type III: budget	15%	10%
Type IV: control	<10%	5%

CONTROL OF ALLOWANCES AND CONTINGENCY

Activity allowances and contingency sums are often bracketed together as 'contingencies' when considering the total amount allocated to cover uncertainties arising within the project scope. Contingencies included as cover for project uncertainties in the 50/50 estimate, which is the basis of the project budget, are controlled by the project manager.

The amount of total contingencies available to the project manager will decrease as the project is executed. If, for example, the budget at contract award includes 10 per cent total contingencies, the rate of decrease with project completion would be roughly as in Table 8.2. The overrun allowance, which is an allowance over and above the 50/50 estimate, is not included in the project control budget. It is included in the corporate budget and is controlled by corporate management.

Table 8.2 Contingency versus contract stage for a total contingency at contract award of 10 per cent

Stage of contract	Contingency
25% complete	7.5%
50% complete	5.5%
75% complete	3.5%
Mechanical completion	1.5%
Commissioning complete	0.5%

8.4 Phasing of costs

When the base estimate of the required type, accuracy and confidence has been completed, the phasing of cost expenditures over time must be determined (step no. 6 in 'cost engineering', Section 8.1). This expenditure/time profile reflects the durations of the project activities such as design, procurement, construction etc. associated with the project breakdown items, and the lead-times imposed by the overall project schedule.

The base estimate is initially made in 'constant value money' (CVM) terms, referred to a specified date. This is the reference date for cost phasing. The phase cost estimate may then be presented either in CVM (stipulating the reference date) or in terms of 'money of the day' (MOD).

MOD is estimated by applying an escalation factor to the CVM estimate. The escalation factor represents a forecast of general inflation plus the market conditions specific to the project. MOD estimates are used if the phased cost estimate is to be used in economic analysis.

8.5 The cost of making cost estimates

Making any cost estimate requires the expenditure of time and effort. Both are always in limited supply, and estimators are always under pressure to produce the most accurate estimates as quickly as possible, with the least possible expenditure of resources (usually engineering manhours, over which the estimator has no control). On the other hand, estimators are often inclined to spend a great deal of time on details of allowances and contingencies, to the detriment of getting better input data and optimizing the project plan. It is therefore important for the project manager – and the project leader – to have a sound 'feel' for the cost of making cost estimates of the various types.

As a rule of thumb, the cost of a project estimate is related to its accuracy roughly as shown in Table 8.3. No estimate can be made precisely accurate. At an accuracy bandwidth of about ±5 per cent, the cost of making the estimate begins to exceed the value of the (estimated) information obtained. This explains estimators' frequent reference to 'the 5 per cent fog' in their best estimates.

Table 8.3 Relationship between the cost of a project estimate and its accuracy

Estimate type	Accuracy bandwidth(±)	Estimate cost as % total project cost
Type I, screening	40%	0.1–0.2
Type II, study	25%	0.2–0.5
Type III, budget	15%	0.5–1.0
Type IV, control	<10%	1.0–5.0+

8.6 Cost estimating systems

Some organizations maintain cost estimating systems, which typically comprise:

- *a manual*, which sets out the methods to be used for preparing estimates of the various types used in the organization. The manual will also set out methods for recording and reporting data for the estimating data base.
- *an estimating data base*, for all estimates other than those made up from confirmed quotations from suppliers. The data base holds historical data on the costs of hardware items and unit cost rates for project functions, with scale, time and location factors. It also contains information for use in cost phasing.

  The data base must be kept up to date by consistent recording of technical and cost information from completed and on-going projects. Otherwise, the data base ceases to be useful as the foundation for reliable estimates for future projects.

Such a system can readily be computerized, when good maintenance becomes even more important. However, even without a computer, the systematic collection of all kinds of cost data is well worth doing. Indeed, in former times, all young engineers were expected to build up their own cost estimating systems from journal cuttings, contacts with suppliers and personal experience – and to keep their files up to date! This is still good practice, especially for those young engineers who aspire to become project leaders or project managers.

8.7 Alternative approaches

While the principles of cost engineering described above are widely accepted, their interpretation and application are influenced by corporate culture, resulting in somewhat different approaches. Consideration of the differences helps to put estimating practice into perspective.

Commonly, as we said, estimates are classified according to purpose and accuracy. Variations in nomenclature, accuracy bandwidth and the treatment of 'confidence' are illustrated by the following examples.

Company A

This company uses the following four estimate types:

- *Order of magnitude*, accuracy bandwidth ±50 per cent; (b)
- *Pre-sanction* (class III), accuracy bandwidth > ±20 per cent;
- *Sanction* (class II) (a), accuracy bandwidth < ±20 per cent;
- *Re-affirmation of sanction* (class I), accuracy bandwidth < ±10 per cent.

The estimate type (a) is the basis for the project control estimate. The accuracy bandwidth denoted (b) is the accuracy bandwidth for 'preliminary order of magnitude' estimates. Intermediate order of magnitude estimates are also used, with accuracy bandwidth between class III and ±50 per cent.

All these estimates have a probability of achievement between 15 and 85 per cent, i.e. 70 per cent confidence that the final project cost will fall within the accuracy bandwidth (cf. the confidence interval of 80 per cent mentioned earlier).

Contingencies for uncertainties within the project scope are estimated on a basis somewhat different from that described above, but overall are typically:

- 15 per cent for conventional projects
- 20 per cent for novel or high technology projects
- 20 per cent for off-sites.

The corporate culture underlying this approach derives from experience of large-scale projects involving contractors in a multinational operating environment.

Company B

This company uses the following four estimate types:

- *Exploratory*, or class D, accuracy range +20 to –40 per cent, average ±30 per cent;
- *Preliminary*, or class C, accuracy range +10 to –20 per cent, average ±20 per cent;
- *Sanction*, or class B, accuracy range +7 to –10 per cent, average ±10 per cent;
- *Control*, or class A, average accuracy range ±5 per cent.

In this approach, probabilistic estimates of confidence are not used. The base estimate is derived from an empirical relationship between (1) a reasonably optimistic assessment of the project cost, and (2) the (higher) cost of an alternative way of achieving the same project objectives.

Contingency add-ons are also based on this relationship. The amount of the percentage contingency is typically similar to the positive accuracy bandwidth quoted above, but tends to be greater with decreasing accuracy.

The corporate culture underlying this approach derives from experience of relatively small, but complex, process industry projects using technologies mainly developed in-house.

Company C

This company uses the following five estimate types:

- *Early screening*, or class V, $P = 1/3$*, contingency 20–25 per cent;
- *Project development*, or class IV, $P = 1/4$, contingency 15–20 per cent;
- *Basic design*, or class III, $P = 1/5$, contingency 10–15 per cent;
- *Design specification, lump sum bid or control estimate*, or class II, where $P = 1/10$, contingency 10 per cent;
- *Detailed design or construction*, or class I, $P = <1/10$, contingency <10 per cent.

The probability denoted by the asterisk (*) has the following meaning: when the estimate includes the stated contingency, it has the stated (fractional) probability P that actual project costs will overrun the estimate by more than 10 per cent.

In this approach, the accuracy bandwidth is constant for all types of estimate, but the probability of exceeding the bandwidth decreases with increasing project definition (and decreasing contingency). Only the positive side of the accuracy bandwidth is considered.

The corporate culture underlying this approach appears to be oriented rather towards construction-type projects, predominantly in the corporation's home territory.

For a given project the different approaches probably only reflect different degrees of corporate pessimism. They do not make much difference, in practice, to the result of the estimate, although they may make a difference to the way the estimate is interpreted.

8.8 Cost control

In cost control, estimates of the appropriate type are used in conjunction with project scheduling to provide the information needed to monitor the cost-effectiveness of work completed and, where necessary, to support project management decisions on corrective action. The amount and extent of technical information available determine the accuracy and therefore the appropriate purpose of the estimate. The following examples summarize the technical information required for the estimate types discussed above. The examples are drawn from process industry practice; other industries have analogous information requirements.

SUMMARY OF TECHNICAL INFORMATION REQUIRED FOR ESTIMATE CLASSIFICATION

Screening estimates

Purpose Project initiation, earliest economic appraisal, preference ranking of alternative projects, development optimization studies.

Summary of technical information required for raw estimate

- General description of project;
- Key project and technology parameters;
- Major hardware breakdown at level of system groups: throughput; capacities;
- Approximate site location, conditions of site terrain, environmental conditions;
- Approximate data on any existing facilities and infrastructure.

Estimating method Factoring.

Feasibility study estimates

Purpose More detailed techno-economic analysis and appraisal of preferred alternative project schemes.

Summary of technical information required for raw estimate As for screening estimate, plus the following:

- additional information from studies and surveys, especially relating to planning and design;
- major hardware items breakdown at level of systems.

Estimating method Factoring.

Budget estimate

Purpose Preparation of project development plan; approval of budget for further front-end engineering (exceptionally, for project execution).

Summary of technical information required for raw estimate As for the feasibility study estimate, plus the following:

- preliminary environmental impact assessment (EIA) and health, safety and environment (HSE) reports;
- details of raw materials, utilities, services requirements;
- operating methods;
- process flow diagrams;
- energy and material balances;
- site maps;
- major equipment lists;
- construction methods;
- preliminary network schedule and barchart programme.

Estimating method Factoring plus some quantities (e.g. major equipment) from materials take-off (MTO).

Control estimate

Purpose Preparation of project implementation plan; checking contractor bids; cost reporting and control; measurement of changes; performance appraisal; forecasting of trends; budget revisions; project reporting; project debriefing.

Summary of technical information required for raw estimate As for the budget estimate, plus the following:

- design philosophy
- HAZOP and risk assessment
- EIA
- design basis
- detailed scope of project work
- project specification

- project schedule
- site surveys
- system specifications
- plot plan and equipment layout
- equipment specifications
- piping and instrumentation diagrams; electrical single-line diagrams
- procurement plan
- contracting plan
- network scheduling
- barchart programme.

Estimating method Complete MTO.

BUDGETS

A budget is an amount of money approved for expenditure on a particular defined scope of project work. The corporate allocation of funds to projects (and to other corporate business) is decided on the basis of budgets. A budget is also a yardstick against which the project's need for resources is assessed and project management performance is measured.

Budgets must be prepared at various times during the evolution of the project. They will be required for each of the main phases – feasibility, definition and implementation – as well as for the total cost of the project. Budgets are also required for funds to make up any shortfall in a previous budget, and for budget revisions and updates. In the case of very large projects, where certain project activities continue for a long time, the *annual* cost of, for example, engineering may be separately budgeted.

All project budgets are prepared in the form of proposals from the project manager to the budget approving authority (e.g. the board of directors or a nominated executive). The proposal enters an approval cycle, defined by corporate policy, to establish whether it is acceptable, by which level of authority, and under what conditions. Like approvals in general, the approval of project budgets is facilitated by standardizing the budget proposal documentation.

In large organizations, project budgets (which include allowances and contingencies against uncertainty) must be integrated with the corporate budget system, because it is the latter which determines the details of project budget submission and reporting.

From time to time, project budgets may have to be revised. Because the budget is important as an instrument of project control, any revision should be undertaken with care. Revisions can be justified (1) if a change in the project scope of work has been approved, or (2) if the most recent cost forecast exceeds the current approved budget. Budget revisions must be proposed and approved in the same way as the original budgets.

8.9 Cost estimating: basic methods

OPERATING COST ESTIMATES

Operating costs, other than the components related to charges on capital investment, are usually derived from the organization's previous experience of operating similar facilities, or are based on information about the performance of processes and plant or equipment supplied by licensors of technology and/or contractors. This information is typically described as 'expected' performance data. Subsequently, through negotiation, the information will be firmed up and presented as 'guaranteed' performance data. The acquisition of reliable performance information is sometimes expensive (see Chapter 11).

The estimate for capital charges (e.g. depreciation, insurances etc.) is typically decided by corporate financial managers, usually as an annual percentage of the project capital investment. Maintenance cost estimates are sometimes decided in the same way. However, using an 'average' annual percentage of the capital investment can be misleading, because maintenance costs for some projects (e.g. those working under severe operating conditions, corrosion, etc.) tend to increase rapidly after a few years operation.

Estimating working capital requirements is similarly problematic. Net working capital is defined as:

current assets – current liabilities

Current assets typically include cash items, accounts receivable, inventories at various stages of completion and prepaid expenses. Current liabilities include notes and accounts payable, and accrued liabilities. Typically, the dividing line between current and non-current is 12 months.

Aside from cash items, notes payable and current maturities in long-term debt, the required investment in working capital depends on customer and supplier credit policies, and the duration of the production and inventory cycle. This is because companies usually aim to buy raw materials and hire labour etc. on credit, while they extend credit to their customers. (This can easily get out of hand. Standard terms of sale which require payment of invoices within 30 days can be eased out to 90 days under competitive marketing pressure, with serious consequences for working capital requirements.)

Working capital is part of the permanent investment base due to its revolving, ongoing character. Because of its nearness to cash, it is also a one-time source of funds. (Accounts receivable can be used as collateral, or sold, as can the element of inventories due to finished goods.) But if working capital is used in this way, it is essential to have alternative sources of funds available at acceptable cost.

CAPITAL COST ESTIMATES: INTRODUCTION

Basically, there are only two methods for making capital cost estimates: (1) based on materials take-offs (MTOs), and (2) based on historical data and factors to account for change. Both kinds produce cost estimates which are derived from a model of the project hardware to which a range of functional activities (i.e. 'design', 'purchase', 'construction', etc.) have been applied.

The success of the estimate depends on how closely the model resembles the eventual project in terms of scope, complexity and completeness of detail. In addition, the estimate will include allowances, contingencies and other costs which depend on the circumstances of the project. Remember that each project is unique. It follows that the basis of the estimate, i.e. its purpose and required accuracy, the scope of the project and its interfaces with existing facilities and with the outside world, and the conditions under which the project is expected to be undertaken, must be clearly understood and agreed between the estimator and the project leadership before the estimate is made.

ESTIMATES BASED ON MTOs

These require extensive project definition and specification of hardware and its associated functional activities, so that the estimate can be more or less fully based on vendor quotations (e.g. 'budget' and 'control' estimates; see above).

In the case of budget estimates, quotations should be obtained for major items of process equipment. The costs of other items, bulk materials and functional activities can be derived by the use of factors, based on firm layouts, flowsheets and similar design data. In the case of control estimates, quotations should be obtained for all the main items of equipment. Other costs are derived from MTOs based on frozen designs, using an up-to-date materials cost data base. This amounts to a 'shopping list' of all the equipment, materials and functional activities needed for the project, and should – subject to allowances and contingencies – provide the most accurate cost estimate.

To help ensure that nothing is omitted from the 'shopping list', checklists are available. These may be obtained from specialist professional organizations such as the Association of Cost Engineers, or under joint venture agreements with some industry corporations. It is wise, however, to build up one's own checklists, if at all possible.

ESTIMATES BASED ON 'FACTORING'

For project screening and feasibility studies, and for some components of budget estimates, the expense of working up MTOs cannot be justified, and the estimates are based on historical data rather than on vendor or supplier quotations. Factors are applied to the historical data to adapt them to the requirements of the project in hand.

The information needed for cost estimating based on factoring includes the following:

- **historical reference information,** which must include the date and location for which it is valid;
- **a time factor** to correct for price changes between the date of the reference project and the date of the project in hand;
- **a scale factor** relating the size (e.g. nominal capacity, throughput) of the reference to the size of the project in hand;
- **a complexity (or completeness) factor** relating the project's technical complexity or completeness to that of the reference;
- **a location factor** to account for any difference between the project site and the location of the reference project;
- **currency exchange rate** adjustment if the project in hand is denominated in currencies different from the reference;
- **contingency, allowances** and other costs to adapt the reference information to the circumstances of the project in hand.

Historical reference data

The reference project should bear the closest possible resemblance (in function, performance, scale, date and location) to the proposed project. The greater the differences between the reference and the project in hand, the greater the likelihood of introducing significant errors into the estimate. This applies to individual items of equipment as well as to a complete installation.

Time factor

This is a multiplier which accounts for changes over time (e.g. due to inflation and/or market forces) in the cost of identical items at the same location. Time factors are occasionally published in the specialist journals, but are often composite indices of dubious origin and should be used with caution. Large companies which develop many projects maintain their own tables of time factors for in-house use.

Scale factor

This is a multiplier representing the change in cost of similar projects in relation to a change in nominal capacity. The well-known scale factor used in the process industries is the so-called 'two-thirds rule' (sometimes reduced to 'six-tenths'), i.e.

$$\text{cost(B)} = \text{cost(A)} \times [\text{capacity(B)}/\text{capacity(A)}]^{2/3}$$

where A refers to the reference project and B refers to the proposed project. The 'rule' presumably derives from the notion that, because much process equipment is like a pot (or kettle), capacity can be represented by a function of volume (i.e. linear dimension *cubed*), while cost is a function of the area of material needed to keep that volume in place (i.e. linear dimension *squared*) (*see* Fig. 8.4). The 'rule' is obviously unreliable if the configuration, process conditions or materials of construction of the proposed project are significantly different from the reference. It is also unreliable when the project capacities differ by a factor of more than about 2 or 3.

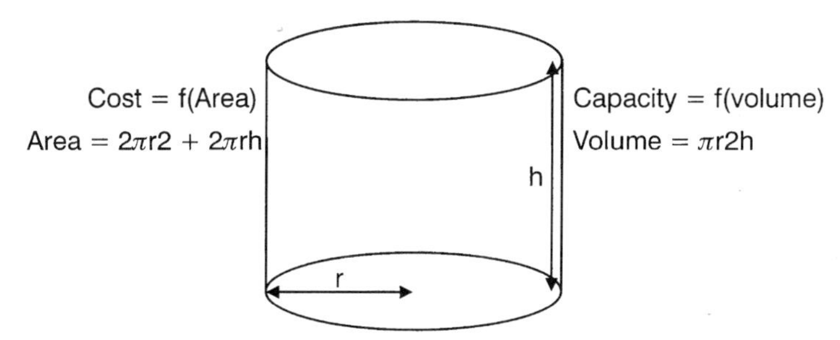

Fig. 8.4 Cost-capacity relationship. Area is a (linear dimension)^2 function and volume is a (linear dimension)^3 function.

In practice, the scale-up exponent is found to vary from about 0.35 to about 0.9 for various process types and for various types of equipment. Values for plant and equipment scale-up exponents are published from time to time in the specialist literature, often in the form of graphs on log paper relating cost to a capacity or performance parameter, when the relationship is typically a straight-line one within limits. The relationship should not be extrapolated beyond the given limits.

Complexity/completeness factor

In estimating for 'complete' facilities, the estimator must consider whether the project in hand includes, or implies, more or fewer unit operations than the reference project. For example, in a 'complete' oil refinery, does the reference include steam generation or power recovery?

For some types of installation, complexity factors can be found in the specialist literature. In general, however, the estimator will have to rely on the project manager for guidance as to what is to be included (and what left out, *and so specified*) from the estimate.

Location factor

This is a multiplier correcting for cost differences between the location of the reference project and the proposed location of the project in hand. It is a composite made up from estimates of local productivity and materials/equipment costs. The international contractors and the multinational operating companies maintain lists of location factors for in-house use, and data are sometimes published in specialist journals. Data are usually available only for countries where there is considerable project activity.

Local market factors

From time to time, project activity in a region may be sufficient to distort the market for some relatively scarce engineering skills or capabilities, or for critical equipment or materials. In such cases, market forces also affect local project costs, tending to force price increases which may be significant. (Conversely, a shortage of project work tends to push prices down.) Some companies track project-related market forces carefully, for in-house use. Valid published data are rare.

Exchange rates

These apply only if the project in hand involves currencies different from the reference. The actual expenditures for the project in hand will take place in the future, so that *forecast* exchange rates must be used. The estimator should obtain specialist advice in this area.

Allowances and contingency

Methods of estimating and applying allowances and contingencies are set out in the preceding section. Note, however, that reference project data from published sources usually include unspecified contingency and allowances. Reference data from contractors are also likely to include unspecified amounts of contingency and allowances, depending on the contract format for the reference project. Contractors are not likely to divulge this information. In particular, beware of contractor estimates quoted on an 'overnight' basis: this assumes zero inflation during implementation of the project, and can be misleading.

Owner organizations normally incur some project costs over and above the contractor's quotation, even on a lump sum basis. Owners' costs include such items as site supervision, provision of temporary facilities, administration, management fees and services, commissioning and insurance. They tend to be lower (as a percentage) for large projects, and higher for small projects. Allowance should be made for owners' costs, although at the initial study stage little information will be available and the estimate will have to be made on the basis of previous project experience.

Inclusions and exclusions

Any estimate must be based on a clear statement of what it covers. There are two ways of developing the statement of what is covered in an estimate:

1. **List of what is included.** This is typically provided by a contractor when bidding for a project, on the principle that anything not mentioned will not be supplied by the contractor.
2. **List of what is excluded.** This is typically in the mind of the owner's estimator, as a catalogue of items which will be supplied by someone else.

While the sum of '1' + '2' *should* comprise the total scope of supplies to the project, any item which has been omitted falls to some unspecified party and will inevitably cause problems when the need for it is noticed (usually well after contracts have been placed!). Good project management ensures that the lists are kept up to date, are properly complementary, and that nothing necessary has been omitted.

OTHER METHODS FOR APPROXIMATE ESTIMATES

Process step scoring

Each step in the process is awarded a 'score' based on, for example, throughput, operating temperature and pressure, reaction type, corrosion problems etc. The capital cost of the estimate is then estimated from an empirical relationship between the total score for the combined process steps and the total capital cost. The method has mainly been used for chemical plant projects, and although it has been described in the technical literature, detailed data have not been published.

'Roll-up' factors

Also known as 'Lang' factors, these are a set of multipliers for correcting equipment item costs (obtained from vendors) to net installed plant costs. Again, detailed data will be available only from corporate experience.

Somewhat similar methods, based on the use of combined rates for functional activities applied to major equipment items to build up the total project costs, also depend on the availability of a sufficiently comprehensive accurate and up-to-date cost data base.

Computer-based estimating

Several computer-driven estimating systems are commercially available. The necessary data bases contain information on equipment and materials costs, rates for functional activities and various on-cost factors, typically presented in the form of a user-friendly checklist. The data base must, however, be customized by the user, who must also take care to keep the information up to date.

8.10 Summary

Cost estimates are vital for project development and for project control. They are prepared with a particular purpose in view – e.g. for screening project concepts, for studying economic feasibility, for budgeting, for managing project implementation – and should not be used for other purposes than the one for which the estimate has been prepared, because it is the particular purpose which decides the level of accuracy of the estimate.

The accuracy of the estimate follows from the amount of technical information available. Greater accuracy always requires more effort to prepare technical information. The cost of this increases very rapidly in relation to the level of accuracy achievable, so that it is always tempting to use preliminary estimates (such as those based on factoring) for higher level purposes, where estimates should be based on more detailed information (such as MTOs). Even when detailed information is available, it is seldom cost-effective to aim at better than ±5 per cent accuracy.

Estimates always include some allowance (contingency) to cover unknown or uncertain project costs. Corporate practice determines how much of this money is at the disposal of the project manager without reference to higher authority.

9

Project Appraisal and Profitability

9.1 Introduction

Investors, be they agencies of government, corporations or individuals, commit their resources to a project on the basis of forecasts (Chapter 2) and estimates (Chapter 8). These are the bases of cost/benefit analysis which, for a growing proportion of projects, means appraisal of profitability.

The appraisal of project profitability has always been important for investors. It is getting more important now, because:

- the scale of project financial commitments is increasing;
- large projects affect the lives of more people now than was previously the case;
- international competition is more severe;
- projects are more complex, with less opportunity to make changes easily;
- mistakes are more dangerous.

The aims and methods of project appraisal may be summarized in general terms as follows:

- **Primary objectives**
  - to estimate the outcome of projects before committing significant funds;
  - to compare with the estimated outcomes of alternative investments;
  - to compare forecast returns with the estimated cost of financing;
  - to assess the risk that the project may fail to come up to expectations.
- **Primary methods**
  - marketing studies
  - techno-economic studies
  - feasibility, viability, strategy studies.
- **Main techniques**
  - cash flow estimates
  - rate of return or 'earning power' estimates
  - sensitivity and risk analysis.

9.2 Cash flows

Projects come into existence because their sponsors or stakeholders believe that a particular project will create more 'benefit' than it will cost. For some kinds of project, this is not straightforward, even when the project meets the conventional criteria for 'evaluability'.

We can, for example, readily visualize a project for which the primary objective is safety, complying with legislation, welfare or operational necessity. Except for the latter, the project generates no cash income (although it may be possible to impute a 'shadow' cash value to project outputs). And in the last example, where operational necessity includes, for example, replacing essential equipment, or where the consequences of not carrying out the project exceed the project cost, the cash flows generated by the project are absorbed in other more general operations. These are examples of projects which do not have quantifiable cash flows (or, to be more precise, cash flows which can be quantified on reliably comparable bases).

However, for commercial projects, cost/benefit analysis follows a fairly standard procedure. We have to compare 'cost' and 'benefit' expressed as flows through time of cash* money. The 'profitability' of a project is its potential for generating significant cash (or cash-related) benefits relative to its costs throughout its lifetime.

Consider the 'benefit' of an oil or gas field. It is based directly on the production profile of the field development. The volume of oil (or gas) flowing each year – we typically take annual periods in calculating cash flows – is the basis of cash inflows to the project, derived from sales. That is to say, cash inflows begin some time after the beginning of the project and resemble the production profile of the field.

Now consider the 'cost'. It has two components of prime interest to us:

1. the capital expenditure required to develop and realize the project (CAPEX);
2. the operating costs required to run and maintain it (OPEX).

On a year-by-year basis, CAPEX rises to a peak, then falls off when building the facilities is complete, although additional capital expenditures may be incurred in subsequent years by modifications, revamps, de-bottlenecking etc. On a year-by-year basis, OPEX might be expected to be fairly constant. It tends, however, to increase due to higher maintenance costs as the equipment gets older, and the effects of stricter HSE requirements. Figure 9.1 shows these trends.

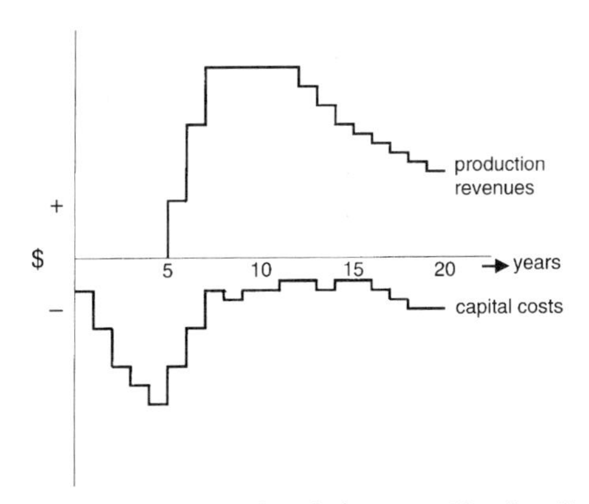

Fig. 9.1 Production revenues and capital cost profiles for oil production.

In addition to these costs, we must take account of taxes and other forms of payment due to the Government. The amount and the timing of these expenses depend on the tax legislation applicable to the project. For example, in some fiscal systems, CAPEX is held to purchase assets and the assets may be depreciated through tax allowances, while OPEX is considered to be absolute expenditure which can be deducted before tax is applied.

* Cash money = money at call or short notice, i.e. actual notes or coins or quickly realizable deposits and securities.

We can now calculate, for each year of the project's life, the annual net cash flow. It is the difference between the gross annual income earned by the project, and the sum of all the annual expenditures. To recapitulate:

- **gross income:** expected output of products (multiplied by) estimated future selling price;
- **expenditures:** CAPEX, OPEX, taxes;
- **net annual cash flow** = annual gross income – annual total expenditures.

9.3 Discounted cash flows

Many projects have long lifetimes, often 20 years or more. The value of money changes through time; usually money loses value through the effects of inflation. For example, 100 dollars received in 1, 10 or 20 years time would be worth less to you than 100 dollars received now. In other words, we have to *discount* future receipts in order to be able to compare them in terms of today's *present* value.

Thus if C_t is a cash flow received in year t, and f is the annual rate of inflation, the present value of C discounted for inflation is

$$C_t/(1+f)^t$$

The net annual cash flow calculated above is expressed in terms of 'money of the day' (MOD). When discounted for inflation, money has a constant value, i.e. the 'discounted-for-inflation' cash flows are in 'real terms' (RT).

Now, however, we have to remember that the project is an investment. It is expected to earn a return, i.e. the total discounted (RT) cash flow is expected to be positive. Indeed, the project is only one of many investment opportunities open to us – we could spend the project CAPEX on something else. We therefore need to see whether the project is likely to be a reasonable way to invest our money in comparison with other opportunities foregone when we invest in the project.

One way to look at this is to recognize that money in the company's treasury is invested and earns interest. We will call the rate at which it earns interest the company reinvestment rate. Clearly, any project which the company considers investing in must earn at least the company reinvestment rate of interest, so we could use this rate (i) as the discount factor in a second phase of discounting. The present value of an annual cash flow C_t is then

$$C_t/(1+i)^t(1+f)^t$$

At any other than very small values of i and f, annual cash flows can be seen to shrink quite dramatically with increasing time in the project's life.

This emphasizes the significance of those cash inflows which occur early rather than late in the project cycle. It often seems to be worth spending quite a lot of money to fast-track a project so as to get product early to the market, and it can be disastrous if projects are much delayed.

9.4 Profitability indicators

When we consider the merits of any project proposal (remembering that the project concept has come into existence because we want to bring about advantageous change in our business performance), we inevitably think about comparing *that* project with some alternative way of using our money. In order to make the comparison, we need to construct a measure of the potential project's estimated profitability: a profitability 'indicator'.

There are several such profitability indicators or criteria. None can be regarded as a totally reliable indicator by itself, so project profitability is usually assessed with reference to several criteria. Some of the more frequently used ones are discussed below.

PAYBACK TIME

A very simple and straightforward way to estimate the profitability of a project is to work out how long it would take for the project to earn the amount of the investment. This is the 'payback time'. It is a satisfactory indicator for short-term investments, when the value of money does not change significantly during the life of the project, and for situations where there are plenty of similar examples to use as references:

$$\text{Payback time} = \frac{\text{original investment}}{\text{total earnings/project life}}$$

RATE OF RETURN (PERCENTAGE ANNUAL RETURN)

This is derived from the payback time, and has similar limitations.

$$\text{Percentage annual return} = \frac{\text{total earnings/project life} \times 100}{\text{original investment}}$$

However, for long-term, specialized investments (such as major industry projects) the change in the value of money over time can be very significant, and must be taken into account when assessing project profitability. For this reason, discounted cash flow (DCF) criteria are generally used as profitability criteria for these projects.

Both payback time and rate of return may be discounted to take account of the time value of money. However, the results of applying DCF will be considerably affected by the choice of the discount rate. Indeed, the discount rate may itself be used as a criterion of profitability.

DISCOUNT RATES

Inflation rate

When we apply a discount rate for inflation to the net cash flows of a project, we often assume that the same rate of inflation applies equally to cash inflows and cash outflows. In fact, this is seldom the case, though the difference is not very significant when inflation is low and relatively steady. In times of high or fluctuating inflation, it is wise to make allowance for the effects of different rates of inflation on inflows and outflows.

Company reinvestment rate

This is a forecast of what the company expects to receive from the reinvestment of profits. The forecast should be available to people assessing project profitability, and the project assessors need to be sure that the company's accounting system has produced a reasonably accurate picture of the company's reinvestment rate. One guideline is to consider the average profits over recent years in relation to the average capital employed over the same period – data which are normally found in the company's end-of-year financial reports. We need to remember, however, that company financial reports are produced with aims other than project appraisal.

Hurdle rate

The hurdle rate is set equal to, or higher than, the company's true reinvestment rate, so as to ensure that the company's profitability is maintained or improved as a consequence of the project. The hurdle rate should not be set according to precedent, or to compensate for notional risks (which are accounted for elsewhere in the project appraisal), but as the rate required to achieve a target rate of return over a specified period of years.

DCF CRITERIA

Net present value (NPV)

This is the sum of all the annual (RT) cash flows discounted at a specified discount rate, such as the company hurdle rate, throughout the life of the project. It is expressed as a sum of money. If the NPV is positive, the project is considered profitable. If two projects are compared, the project with the larger NPV is considered the more profitable.

Internal rate of return (IRR)

This is the discount rate at which NPV = 0. It is expressed as a percentage. If the IRR is greater than the hurdle rate, the project is considered attractive. If two projects are compared, the project with the higher IRR is the more attractive. The minimum requirement for developing a project would be that the project IRR is greater than the company's reinvestment rate.

Growth rate of return (GRR)

This is the rate of interest at which the project capital would have to be invested in order to amount to the total of project annual incomes invested at the company reinvestment rate. (By contrast, IRR is the rate of interest at which project capital would have to be invested in order to amount to the total of project annual incomes invested at the same rate.) The higher the GRR, the more attractive the project.

Notice that none of these DCF criteria takes account of the fact that capital for investment in projects is limited. But no company has infinite capital resources. The following methods help one to choose the most profitable projects when total investment is limited and take explicit account of project investment requirements.

Maximum capital exposure (MCE)

This is the trough of the cash flow profile, i.e. the greatest amount of money which will be committed to the project at any one time. If the MCE is too high, the company may reject a project even though it is attractive in terms of NPV, IRR or GRR.

Discounted profit/investment ratio (DPIR)

This is

$$\frac{\text{NPV of project @ company reinvestment rate}}{\text{NPV of capital investment @ company reinvestment rate}}$$

The greater the DPIR, the more attractive the project.

Capital productivity index (CPI)

This is:

$$\frac{\text{NPV of cash incomes}}{\text{NPV of capital investment}}$$

using the company reinvestment rate for discounting, i.e.

$$\text{CPI} = \text{DPIR} + 1$$

The higher the CPI, the more attractive the project. CPI is *income* per unit of investment, while DPIR is *profit* per unit of investment.

9.5 Profitability analysis

Profitability analysis is a technique for modelling a project so as to investigate its economic performance. The technique is used, for example, to:

- estimate the economic feasibility of a project considered as an investment opportunity;
- rank several projects in terms of their relative attractiveness as investment opportunities;
- provide evidence on the value of an asset (such as a production facility) as collateral for debt;
- establish a basis for pricing when buying or selling an income-generating asset.

The magnitude, time-scales and uniqueness of major industry projects are such that profitability analysis relies almost wholly on DCF models. But, like any other model, a DCF model invokes a distortion of reality. It does so by reason of assumptions, implicit in the construction of the model, which may not be clear to those using the model for making decisions. Such assumptions then become pitfalls, for example:

- Cash flows are assumed to be received at the end of the year, and are discounted discretely on an annual basis. This is merely a convention. Different conventions of, for example, length of accounting period, or discounting basis (e.g. continuous discounting) lead to different DCF results, so the convention used should be made explicit.
- The discount rate is assumed to be constant throughout the project life. There is no reason why it should be, and indeed in some circumstances, this is not a good assumption even for a crude model.

DCF models used for establishing values for borrowing money or for the purchase or sale of an asset are usually based on NPV modified in some more or less arbitrary way to reflect the state of the money market at the time. For example, the NPV is calculated at a discount rate equivalent to the market interest rate (which reflects, among other things, current demand for assets of the kind in question, market confidence, etc.).

In evaluating projects as investment opportunities, DCF models are used in a much more 'pure' form, based on NPV, IRR or their derivative ratios such as DPIR and CPI. In the interpretation of these models, the maximum capital exposure (MCE) should be taken into account. For stand-alone projects, the project manager will also want to satisfy himself that, irrespective of the overall cash flow, NPV, etc., there is enough money available in each project accounting period (e.g. 'milestone') to service that period's debt. Otherwise, his project will stop dead!

Should we work with NPV or IRR? IRR has the advantage of being directly comparable with the cost of money, and with bankers' and businessmen's way of looking at returns, which is more comfortably expressed as a percentage. And since both methods use the same input data in the same way, we might expect that they would rank projects in the same order. Alas, they do not – or at least not always! For example, if project A's earnings are achieved earlier than project B's, A will be less sensitive to the discount rate than B. So A will look increasingly attractive relative to B as the discount rate increases. But, at a discount rate less than the point of indifference, B will be the more attractive.

In other words, the pattern of the cash flows has a bearing on the methodology we choose for evaluating them. Here, there is a practical disadvantage in using IRR. The expression for calculating IRR is a polynomial and has multiple roots. It cannot be solved explicitly. In some circumstances, for example when negative cash flows follow a series of positive cash flows, ambiguous IRR results are obtained. Furthermore, since IRRs are rates of return on varying levels of investment, they are not additive (NPVs are), and so should not be applied to multiple project selection. Essentially, however, NPV measures the surplus earned by the project over and above the opportunity cost of capital (no matter how we define that). IRR measures an abstraction which, if it contradicts what we learn from NPV, has to be regarded as misleading.

Evaluating projects as investment opportunities is always a matter of comparison – either we are comparing one project with another (which may be the 'project' of doing nothing, when we want to establish that doing something is in fact the better option), or we are compiling a portfolio of projects which have to be ranked in order of preference as investment opportunities.

MUTUALLY EXCLUSIVE ALTERNATIVES

In this case, when we must choose between one project and another, we have to consider the earnings of the *difference* between them. For example, suppose we have two projects (A and B) and we can only afford to go ahead with one. Then the rate of return on the difference (B–A) is a measure of the attractiveness of B relative to A.

If there are more than two mutually exclusive alternatives, they are compared in pairs in order to determine the more attractive project. Remember that one alternative is always to do nothing.

MUTUALLY EXCLUSIVE ALTERNATIVES WITH DIFFERENT LIFE SPANS

The present values of projects which have different project lives are usually considered not to be comparable. The difficulty arises because of the implied commitment of earnings from one project to the other, as shown by the rate of return on the difference between the cash flows of the two alternatives.

One method of overcoming this difficulty is to suppose that each project is replicated over a period equal to the lowest common multiple of their lives. This can lead to some extremely tedious arithmetic. However, it can be shown (Karathanassis 1985) that the NPV evaluated to infinity is

$$NPV \times (1 + i)^t/(1 + i)^t - 1$$

The use of infinity as the notional life of the replicated projects provides a common time-base for comparing two mutually exclusive projects. The present value determined in this way is called the 'extended net present value' (ENPV).

PROJECT EXPANSION

This is a special case of comparing the NPVs of two projects, i.e. the project *before* and the project *after* the investment in expansion (revamp or de-bottlenecking). The difference is a measure of the economic viability of the expansion.

FAST-TRACK PROJECT ECONOMICS

Having in mind the effect of discounting on future incomes, we can see that sometimes a situation may arise when it is worth investing simply to earn future incomes sooner (because early future values yield a higher NPV). Fast-tracking a project is a well-known technique which, given suitable planning and resourcing, should be effective. Fast-tracking is, however, vulnerable to 'glitches' – unforeseen disturbances in an expectedly smooth process – and is now regarded with considerable suspicion by many project managers (see Chapter 16).

DISCOUNTING AND 'LIFE CYCLE' COSTS

When a project is treated as a financial investment, it is normal to use a discount rate higher than the interest rate which could be obtained from a fixed-rate, secure investment. This higher discount rate (e.g. the hurdle rate) reflects the 'riskiness' of the project, and may be high enough to carry some important implications for project appraisal.

Discounting reduces the present value of both future revenues and future costs. Typically, revenues arise throughout the life of the project and are expected to conform rather closely to plan. Costs (apart from the investment itself), especially repair and maintenance costs, tend to increase substantially and often unexpectedly in the later years of the project's life. But discounting causes them to appear relatively insignificant. Consequently, the idea of investing, at the stage of initial development of the project, in means of reducing maintenance, repair or close-down costs looks unattractive. For long lifetime projects which may incur potentially large expenditures late in the life cycle, additional investment to minimize such expenditure is difficult to justify when high discount rates are used in the project appraisal – and yet the work will still have to be done.

Because this work is deferred and takes place in the rather distant future, it is not unusual for economic or market changes to force a change in ownership or responsibility in the meantime. Future capacity for undertaking the work is likely to be overestimated. When the work has to be done, the then-available resources will be under pressure, and this inevitably leads to increased costs.

It might therefore be better to appraise this type of project (offshore structures, large petrochemical plants and power stations are examples) on the basis of a criterion other than DCF. This of course requires a change in the basic philosophy of project development, namely, a major project should not be seen simply as a financial investment.

CAPITAL RATIONING: RANKING OF PROJECT ALTERNATIVES

Considering the whole of a corporate investment programme, there will usually be several projects which the company would like to develop during the next few years. But there is seldom enough capital available to invest in all the projects. The company has to choose priorities for projects to develop, i.e. which to defer and which to reject.

The simplest method is to rank the projects in order of, for example, DPIR or CPI. The capital budget limit then fixes the 'cut-off point' beyond which no additional projects can be accepted. In practice, the choice is rarely so easy. For example, a company may have to choose between one attractive but high investment project and several only slightly less attractive projects with perhaps lower overall risk. It is in cases like this that the evaluators will need to consider a range of profitability criteria in making their choice.

EXPECTATION AND RISK

So far, we have based our assessment of profitability on cash flows. Our analytical method has been deterministic, reasonably straightforward and leads to easily understood decision rules. But this approach does not allow explicitly for risk. When various large contingencies and excessively high discount rates are used as surrogates for risk, the method is likely to give misleading results.

All the input data we have used in our analysis so far are in fact merely guesses about the future. There are no facts about the future, only expectations. We *expect* that the investment for a project will be so much and we *expect* that the project will generate income in such-and-such a pattern. But, as we all know from bitter experience, our expectations may not be achieved! In other words, there is a risk that our expectations may not be met, or will not be met as fully as we had hoped.

We define 'expectation' in this context as 'the value achieved if the project (plan) is successful × the probability of success'. This leads us to the concept of expected monetary value (EMV). For EMV at the company reinvestment rate (r) we can say:

$$EMV_r = NPV_r \times \text{probability of achieving } NPV_r$$

or, more generally,

$$EMV = \text{unrisked income (cost)} \times \text{probability of achieving this income (cost)}$$

We now have an efficient decision rule: *'maximize the EMV'*, i.e. choose the projects with the highest EMVs so as to accrue the maximum wealth from the company's investment programme. The evaluator would then (for example) reject projects the IRR of which is less than the hurdle rate, and would choose projects with the highest EMVs based on the company reinvestment rate.

In calculating EMVs, we are using probability explicitly as a surrogate for the risk of not achieving our plan or expectation. People, including project evaluators, have great difficulty with the concept of probability unless there is a well-founded basis in statistics for the case under consideration. In most project situations, this does not exist, so people are inclined to look for 'worst cases', and ask 'what if' questions, e.g. 'What if the project investment increases by 10 per cent?' This is the principle of so-called 'sensitivity analysis'.

The project is tested to see how sensitive the profitability criteria (e.g. NPV, EMV, etc.) are to a range of 'what if' questions addressed to the more significant items of project input data. Examples of cases for sensitivity testing for an oil/gas production project are:

- 90/10 CAPEX estimate
- OPEX plus 25 per cent
- project start-up delayed by 1 year
- production rate/recovery decreased by 20 per cent.

This approach assumes the *certainty* of the test case. The evaluator is in effect saying that if the project remains viable in each of the sensitivity tests, then the argument for accepting it is reasonably robust. The robustness argument is clarified by plotting the sensitivities on a 'spider diagram'. The diagram (Fig. 9.2) helps us to see the influence of each variable on profitability, and the size of the risk of making a loss.

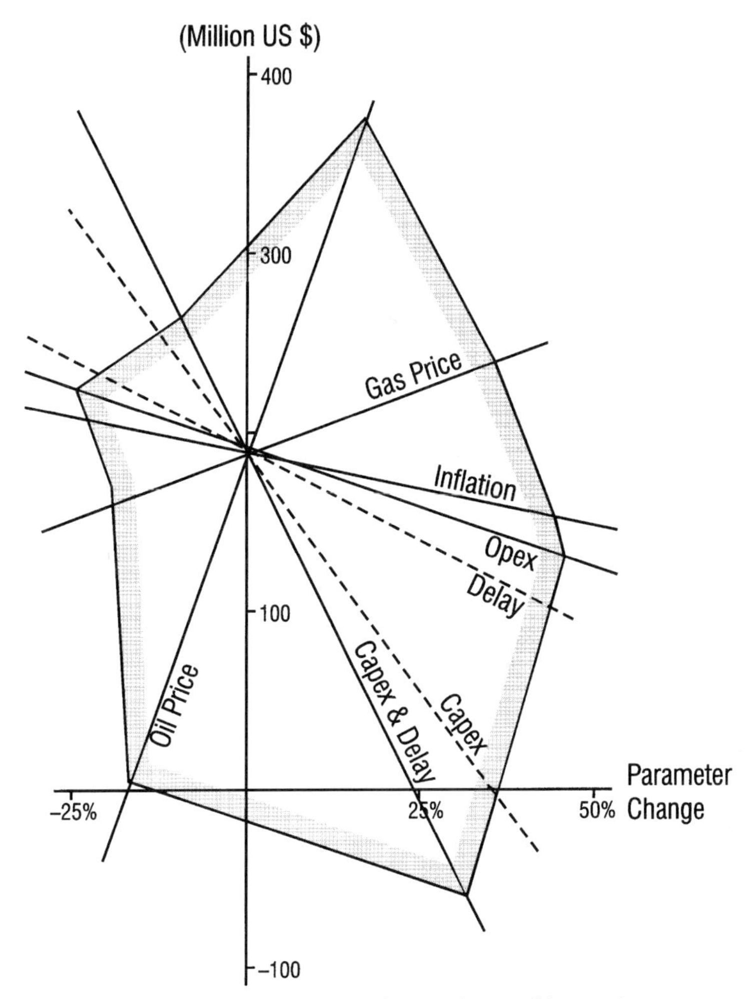

Fig. 9.2 Example of a spider diagram for an oil/gas project.

Assuming the certainty of the test case is a simplification. When we say, 'What if the investment increases by 10 per cent?', we usually mean, 'I think the capital cost could increase by 5 per cent, maybe 15 per cent, but the most likely case would be 10 per cent.' And when we say, 'What if the price of crude oil falls by US$2/bbl?', we may have in mind that it may be anywhere between, for example, US$5/bbl less and US$1/bbl more than its current price. Moreover, looking at each test case situation in a sensitivity analysis may be reassuring, but it doesn't tell us much about the real impact of all probabilities on our expectations for the project. What we are looking for is the *probability distribution* of the profitability indicators of our project.

Probability distributions of project outcomes can be calculated by assigning probability distributions to the project variables (either all of them or, more usually, those to which project outcomes are most sensitive). Unless adequate statistical data are available, these assigned distributions are bound to be subjective. But we will nonetheless get a more useful result from saying, for example, 'The price of crude oil will be US$18/bbl, with minimum US$13/bbl and maximum US$19/bbl' (i.e. a triangular distribution) than by pinning our hopes on the single figure of US$18/bbl.

For most project variables, experience will guide our subjective choice among only a few types of probability distribution:

- **triangular** – minimum, most likely, maximum values;
- **uniform** – minimum and maximum values;
- **normal** – a bell-shaped function completely specified by its mean and standard deviation. It is probably the most frequently encountered random variable distribution in engineering and geological projects.

Calculating project outcomes from the project variables' assigned probability distributions (using a computer to carry out, for example, 'Monte Carlo' simulation) generates a probability distribution of the selected project outcome. This can be interpreted to indicate confidence limits, such as the 'most likely' value of the project outcome, or the value in which we may have '90 per cent confidence'.

This is a powerful method of investigating the riskiness of projects. However, it seems to be less favoured by corporate decision makers than is sensitivity analysis using deterministic data, perhaps because of top management's widespread unfamiliarity with probabilistic concepts.

9.6 Snares and delusions

The techniques of project profitability analysis treat the project simply as a financial investment. In many industries, other factors (e.g. security of supply, policy on depletion of raw material resources, socio-economic and technological development among others) must also be taken into account.

Another reason for not relying exclusively on profitability indicators is this: even though NPV and IRR (say) make the project look attractive, it may fail in the commissioning and early operation phase because:

- this is the time when maximum repayments of debt finance plus interest often fall due;
- this is also the time when start-up and teething troubles arise (NB – especially in the provisioning of spares and other consumables during this period).

Hence we should study the sensitivity of the project during this period (e.g. the first 2 to 3 to 5 years, depending on the nature of the project) separately. The best way is to build models of profit-and-loss accounts and balance sheets in the way that your accountant would do for the actual plant. Note in particular what may happen in the event of severe inflation or devaluation of currencies significant for the project.

Depreciation raises another problem. Depreciation is conventionally not considered to be a cash flow, although it may be allowed by the taxation authorities as a cash operating expense, thus affecting the way it is treated in cash flow calculations. Typically buildings and similar structures are allowed to be depreciated over long periods of time (e.g. 20+ years). Most equipment has to be depreciated over much shorter periods (say 5–8 years, depending on the equipment type and the tax regime). In capital intensive projects, this implies high depreciation charges which may raise the problem of early cash flow deficits. The important point here is that *cash* is not the same as *profit*! From the project's point of view, cash matters more than profit. The project may operate successfully for years generating positive cash flows (and paying its debts), while not generating profits (and consequently not paying dividends).

9.7 High risk situations

INTRODUCTION

A high risk project is one in which there is a higher than average risk that the project will fail. We have to decide what we may consider to be an 'average' risk, and what we mean by 'project failure'. Average risk can be assessed only with reference to the experience of an industry sector – say oil refining or petrochemicals – and, indeed, to subdivisions within the sector, such as FCCs or ethylene cracking. The division is arbitrary, so 'average' risk is what we choose to make it.

Project failure means failure to hit targets such as budget, schedule or performance (e.g. as specified in guarantee tests before acceptance of the project plant). There are other criteria such as the commercial success of the participants in the projects (though it is arguable whether the fact that a contractor makes a profit or a loss has much relevance to the success or failure of the project on which he worked). Criteria, such as those related to social or political aims, are also sometimes taken into account (Baum and Tolbert, 1985), but perhaps the most convenient and accessible definition of project failure is simply 'failure to meet the expectations of the project stakeholders'.

This definition places a large part of the responsibility for ensuring success on those who carry out project studies, for it is in the study phase of project preparation that expectations are investigated and determined, eventually to the point where they form the basis of the project investment decision. Of course, it may be said that sometimes stakeholders' expectations are unreasonable (though, if true, that may also be charged at least partly against those responsible for project preparation).

PROJECT DEFINITION

As propositions, we may say the feasibility of the project should be assessed on an objective basis; that the objectives of the project should be clearly stated; and that changes are likely to lead to overruns of both budget and schedule, and may also adversely affect project performance. We might add that the leadership of a project is effective in so far as it changes value perceptions and increases commitment, thus personalizing 'objectivity'. So the concept of objectivity in project development is easy to challenge, vulnerable to abuse and fragile, but nonetheless important in selling project concepts to investors and their financial backers.

Achieving clarity of objectives is at once very significant and rather elusive. The most successful projects appear always to have very clearly defined objectives, though some projects – especially those which have what are primarily social benefit aims – may suffer from attempts to specify in detail what are inherently fuzzy objectives.

Sometimes there is also a potential mismatch between the objectives of the project stakeholders and the objectives of the project management. An example is the dilemma of project cost against project safety. 'Risks' (I mean here *physical* risks) may be assessed as probabilistically negligible, yet the consequences of an accident may be enormous. Although there may exist considerable general knowledge about engineering reliability and the (physical) risk of failure, new engineering means new risks, with limited opportunity for experimentation, and often no established methodology for making the project decisions which bear on risk management.

The question of changes (i.e. changes in specification affecting the project) has been argued ever since Imhotep built Zoser's pyramid – and probably before that (Fig. 9.3). The argument continues. Is it better (1) to insist that specifications and designs should be frozen at some stage before which the impact of change is unimportant, or (2) to accept that there will be changes, so that what is really needed is an effective way of controlling their impact? The former was a viewpoint more in favour when projects were relatively small, and before the advent of

Fig. 9.3 Imhotep, who built the Step Pyramid at Saqqara for the Pharaoh Zoser (*circa* 2668–2649 BC). He was later deified as a god of wisdom and learning: recognition achieved by few modern project managers. Reproduced with the permission of the Ashmolean Museum, Oxford.

sophisticated project control systems. Some projects do benefit from the imposition of a 'freeze' date. The important factor is that the choice and the decisions are within the control of management, which should recognize that changes arise for two different reasons:

1. necessary but unforeseen changes, such as changes in requirements, or corrective changes;
2. the possibility of additional benefits, i.e. improvements.

In the latter case, the question arises whether the project objectives were properly defined in the first place.

TECHNICAL PROBLEMS

Technical problems arise from uncertainty about technological issues, from the necessity to coordinate complex project interfaces, and from difficulties in design management.

Uncertainty about technology occurs very often in innovative projects. The most extreme example is in research and development (R&D), where project appraisal has much more the character of a monitoring procedure than a means of making decisive choices about investment. Technological uncertainty is made much worse when there is simultaneous pressure of time (so-called 'concurrency'), because the urgent need to show that something is being done inhibits doing it properly (cf. the old engineering proverb 'you can have it tomorrow, or you can have it right!').

Evidently, very much of the most significant technological information will be concerned with design – hence the importance of design management in this context, and the consequent importance of completing sufficient design work in the early stages of the project (but note the problem of design changes mentioned above). These technical problems are not, unfortunately, amenable to algorithmic solutions. Because of their wide-ranging repercussions, they demand that the most experienced and effective management resources be dedicated to their solution if the project is in any way susceptible to technical uncertainty.

FINANCING PROBLEMS

These are to do with the amount of finance required, sources of finance and financial/commercial risk, all of which may be assumed likely to cause project difficulties if they are not adequately addressed at the appraisal stage.

Evidently, the *amount* of finance needed for the project has to be assured both when the project is initiated and during implementation. A frequent difficulty is caused by unforeseen cost increases due to changes in the value of money as well as escalation. Other, less obvious problems arise from the inadequate allowance for working capital. This is an example of difficulties caused by the *pattern* of financial flows, i.e. the availability of finance at the appropriate time.

The sourcing of finance, and in particular the mix of public and private funding, is often especially significant in major international projects, when the provision of financial guarantees may raise questions between financial institutions and sovereign states which lead to delays that eventually affect the pattern of financing on which the project's budget and schedule were planned. The financing of projects is discussed in more detail in Chapter 10.

APPRAISING 'R&D' PROJECTS

As we have seen, for projects where there is a significant delay between making the investment and earning a return, we have to take account of the time value of money in order to be able to compare the cost of the project with its future benefits. We discount the cash flows (C) in each period of the project life (t) using a discount rate (i) to obtain the project's net present value (NPV):

$$\text{NPV} = \sum_{0}^{t} \frac{C_t}{(1+i)^t}$$

Often the discount factor $(1 + i)$ combines the opportunity cost of money for the project with an allowance for inflation and another allowance for the assumed riskiness of the project. This is the way – using a single aggregated discount factor – that most software is programmed, so the single aggregated discount factor gets to be implicit in DCF calculations.

R&D projects are characterized by (1) long intervals between making the investment and earning a return, and (2) successive stages of development, e.g.:

- stage I – technical research
- stage II – market development
- stage III – sales.

The main aim of each stage is to reduce uncertainty, and it is only at stage III that we begin to earn a return.

Inflation

Often, corporate accountants provide a single inflation rate based on company-wide historical information, which is applied over the whole of the project life. It is applied equally to expenditures and income. However, different businesses adjust to inflation at different rates, so applying accountants' practices leads to a mismatch of inflation-adjusted cash outflows and inflows, which usually discriminates against long-term development. Also, the stages of R&D type projects should really be considered as different business segments (and the project as a whole is likely to be considerably different from the company's historical business), so it is unreasonable to apply a single inflation adjustment to the different cash flow segments of an R&D type project.

Risk

The risk coefficient (sometimes called the 'beta' factor) is based on consideration of risk in the industry sector and in the particular project. The beta factor is set at 1 for average risk; at less than 1 for less than average risk; and at more than 1 for more than average risk. Typical industry sector data are available from specialized sources.

Using a single risk coefficient throughout the DCF calculation implies that risks grow geometrically with time, though it is unreasonable to penalize future cash flows for today's uncertainty.

In R&D type projects, we progress from stage I to stage II to stage III only if the preceding stage is successful, i.e. it reduces the uncertainty of the next stage. So although the risk coefficient for stage I will be high ($\beta > 1$), the coefficients for stage II and then stage III should be much lower – otherwise the R&D objectives need to be redefined.

'ENVIRONMENTAL' PROBLEMS

In this category, I include questions to do with the physical environment of the project (geophysical features, location etc.) and social, legal or community questions.

The physical environment of the project is not a variable subject to control: its features must simply be recognized and adequately taken into account in project planning. The most obvious example of the influence of geophysical characteristics' influence on the project is offshore, though there are plenty of onshore locations which pose questions for special consideration, such as design conditions for foundations and plant, access, logistics and communication.

As is well known, the cost of construction at a difficult site may be several times what is normal at ordinary sites where a developed infrastructure already exists. 'Community' questions include topics such as fiscal and regulatory practices and trends. They are not infrequently neglected at the project appraisal stage. Regulatory trends affecting worker health and safety, together with increasingly stringent legislation affecting environmental management, are said to have increased project investment costs in real terms during the past decade, despite improved engineering productivity.

'Community' questions also relate to work practices, the amenity requirements of personnel, recruitment and training, and the attitude of the authorities and the local population towards development of the project. None of these factors is easy to quantify; each one has the potential for pushing the project towards success or failure.

MANAGEMENT AND ORGANIZATION

Clearly, a 'well-managed' project is more likely to succeed, irrespective of how we define success, than a badly managed project. That, however, begs the question of what constitutes good management. An element of good management which is perhaps less subject to circularity in discussion, is 'organization'. At the stage of project study and evaluation, some attention has to be given to the way the project will fit into the owner's organization. The project will, in fact, alter the owner's organization, in ways that should be planned so as to avoid undue stress.

As an example, consider the development of a project to make a new product, from the initial idea to the setting up of the new business centre. The project passes through a series of phases, where the number of opportunities to stop and withdraw from the investment diminish, and withdrawal gets more expensive, as the project proceeds. Organization in the early stages of development is fairly straightforward, and continues to be so up to the final (definitive) study stage, although there may be a need then to make a timely change in project leadership. Project implementation leading to commercialization becomes much more complex, and consequently deserves careful thought at the planning stage, especially when the project is actually one of a series in a major development programme.

Organization of project management is not, of course, immune to the changes in the owner's organization wrought by the project's existence. This too must be carefully planned. Perhaps the most important – and often the most controversial – factor, brought to the forefront in high risk projects, is to provide unambiguously for the project manager's authority commensurate with his responsibility.

THE COST OF RISK AVERSION

So far we have been discussing the risk that the project may fail to come up to its sponsors' expectations as to cost, schedule, performance of its specifications, or any combination of these. In addition, we mentioned physical risk, i.e. hazard to workers' health and safety, and the community at large. In this category, I shall also include environmental hazards. It is obvious that, according to this classification, many industry projects fall into the category of high physical risk projects.

How much is it worth investing to avert these risks? This is a question which is very difficult to answer at the project preparation stage (when the key decisions are taken) because the risk aversion component of project investment is not evaluable in conventional ways. In some cases, one can say that failure to invest in risk aversion will result in the project leadership being sent to jail – perhaps an immediately quantifiable benefit! In other cases, even simple future benefits cannot be realistically measured, and yet such benefits must somehow be balanced against real costs.

A basic problem is that society – which establishes legislative and also customary attitudes to safety, health and environmental questions – views risks differently from the individuals affected. As the perceived probability of a (serious) risk increases, the value of risk avoidance to the individual increases much faster than its value to society (Fig. 9.4). Underlying this problem is, firstly, the fact that an individual's perception of risk is (by definition) subjective, and seldom corresponds with any 'societal' or group view based on statistical models. Secondly, we do not, as individuals, much care for equations that relate the value of *our* lives to sums of money, whatever we may feel about other people's insurance policies. Thirdly, both individuals and society have to find a way of agreeing what level of risk is 'acceptable', especially when the risk is spread over a long period of time. For example, most societies are content to accept the sacrifice of many individuals to exploitation of the motor car, something which is deemed not to be acceptable in the exploitation of nuclear power.

Provided both the risk and the benefits can be assigned comparable (i.e. monetary) values, then some rational approach to a policy for risk aversion can be attempted. The costs of risk reduction

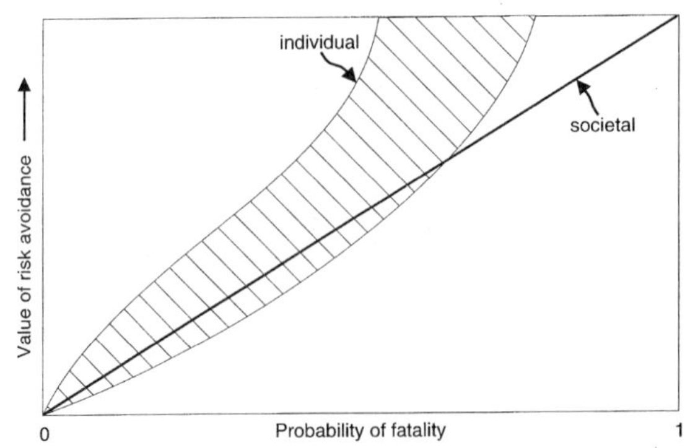

Fig. 9.4 Individual and societal perceptions of the value of risk avoidance (after Starr and Whipple, 1980).

will include investment in, for example, hazard analysis and operability studies, environmental impact assessments, quality assurance, additional monitoring and control instrumentation, additional or improved equipment, and perhaps extended training programmes for workers and managers.

The actual costs of these investments are rarely reported, but – by way of illustration – there seems to be a consensus among specialists in the petroleum industry that they represent some 15–20 per cent additional project cost. About half of this (say 10 per cent) is due to environmental protection legislation, while the remainder is attributed to meeting higher standards of worker safety requirements.

Is this not inconsiderable extra cost worthwhile? Or is it perhaps excessive? We can't answer, because we cannot say what would have been the consequences had it not been incurred. However, given some agreement on the measurement of costs and benefits, a 'cost-effectiveness' approach to risk reduction would be to say that the limit for acceptable risk would be the point where the incremental cost of risk reduction is just balanced by the incremental reduction of risk. This is perhaps a more rational – or at any rate, a more manageable – approach than the balancing of pressure groups which currently determines the 'acceptable' costs of risk aversion.

9.8 Summary

Business projects are typically assessed on the basis of their potential profitability, although many business projects include elements which cannot be evaluated in a straightforward quantitative way.

For most commercial projects, appraisal must take account of the time value of money. The evaluator's view of potential profitability, especially when comparing alternative projects, will be affected by the pattern of project cash flows through time. Discounting may then lead to an apparent undervaluation of cash flows which arise in the long-term future.

The choice of the preferred project option usually demands consideration of more than one profitability criterion, together with testing of the project's sensitivity to a range of hypothetical situations, so as to model the riskiness of the decision to invest (or not) in the project. Often, profitability modelling generates so much information that reduction of data to a form readily assimilable for decision making becomes a key issue. In this, the clarification of all underlying assumptions (and mind-sets) is particularly important.

10
Financing Projects

10.1 Introduction

Small projects are usually financed entirely from corporate funds. If you are responsible for such a project, your main concern will be to keep the company's accountants happy. In large projects, very substantial sums of money obtained from sources outside the company are put at risk in the form of finance for the project. Because of the risk, arranging the finance can be complicated and expensive. The business of negotiating financial arrangements is a main concern of project leaders. Project managers may not be involved at all. But, in that case, they will be unable to contribute their experience to influencing the financing, and will be ignorant of the constraints that financing may impose on the manageability of the project.

Consequently, although they will seldom get involved in the details, project managers, as well as project leaders, do need to contribute to the development of project financial strategy, because it is so important for project success.

Our discussion in this Chapter is based on the financing of major projects. This enables us to consider issues which arise in all projects, albeit with varying emphasis. Major projects have some characteristic features that colour the approach to financing:

- The projects are usually very large, whatever the yardstick used for comparison.
- Their functions – whatever they exist to do – often overflow national boundaries. The impact of a major project as a whole is often international.
- They often exceed the capacity of any single organization to plan, supply and construct, which implies coordination, sometimes internationally, of many different project activities.
- They are technically complex, demanding resources of skill, manufacturing and production which are not widely available. Some projects require items of equipment or materials for which sources of supply are very limited.
- They are often dedicated to a single purpose. A good example is a liquified natural gas (LNG) project, which comprises gas collection, liquefaction, storage, delivery to the market by a fleet of specialized vessels, reception facilities, more storage, regasification and distribution facilities. None of the constituent parts of the overall project can be used for any other purpose without significant additional investment.
- They are often located at remote sites, demanding substantial additional investment in infrastructural work which is not in itself productive.
- They are highly capital intensive.

- The time for project development and implementation is quite long, so that return on the investment is deferred for some years after it is committed.

The planning of finance and the eventual financial structure of the project will reflect these features. So it is hardly surprising that the money men – the suppliers of finance – will be very much concerned to analyse the risks associated with the project before they accept the investment opportunity which it represents.

10.2 Characteristics of a financing strategy

In addition to the typical project characteristics mentioned above, each project will have its own special features which will affect the design of its particular financing strategy. Many such features must be taken into account. However, there are four which have rather wide general application:

1. **The project will fail, no matter what its technical merit, unless enough finance is available to complete it.**
   This problem has been well known for a very long time (Luke XIV, 28–30):

   For which of you, intending to build a tower, sitteth not down first, and counteth the cost, whether he have sufficient to finish it?
   Lest haply, after he hath laid the foundation, and is not able to finish it, all that behold it begin to mock him,
   Saying, 'This man began to build, and was not able to finish.'

   For example, consider the three projects in Table 10.1.

   Table 10.1 Can we really afford the best project?

	Project		
	A	B	C
Technical evaluation (%)	100	95	90
Finance availability (%)	90	95	100

   Which project can go ahead? As things stand, only C!
2. **The design, implementation and management of the financing demands the same level of commitment of planning and management as the project itself.**
   For example, consider a project in which the project cost (say, 100 units) is financed by credit over 10 years at 8 per cent p.a. interest. Now, there is a rule of thumb which says that, in such a project, the cost of finance is approximately as follows:

   - for the first 4 years, 100 per cent of the amount outstanding for 2 years;
   - for the remaining 6 years, 100 per cent of the amount outstanding for 3 years;

   i.e. 100 units of money $\times$ (8 per cent $\times$ 2) + 84 units $\times$ (8 per cent $\times$ 3) = 36 per cent, which is more than one-third of the project cost! Bearing in mind that the total cost of finance may amount to the same order of magnitude as the (unfinanced) project cost, it is clear that good *finance* management is as important as good *project* management.
3. **Financial planning should begin at the same time, or earlier, than technical project planning.**
   Regrettably, financial planning is often begun at the last minute, when most of its other features are already 'cast in concrete'. While it is seldom possible to find the best financial

arrangement without taking account of the special features of the project in question, financial planning need not wait on project detail.

4. **While the financial package is likely to reflect the complexity of the project, finance has some inherent characteristics which themselves add to the complexity of the undertaking.**

   It is not always clear which comes first: identifying sources of supply for finance, or preparing technical specifications for equipment and services etc. Finance, or more often lack of it, imposes the need to develop a sense of compromise. The availability of finance, and the terms on which it is available for a particular project, may be subject to significant and rapid change for reasons quite beyond the control of the people responsible for the project.

   The structuring, and eventually the acceptability, of the financial package depends largely on one's point of view. Obviously, a lender is likely to have a different point of view, at least initially, from a would-be borrower. Moreover, a package which appears attractive to, say, the project contractor who wants to win an order, may not be so attractive to the project owner or his Government authorities (or vice-versa).

10.3 Identifying sources of finance

Without finance there will be no project. Yet many projects around the world are begun and not completed (or run into long delays and huge cost escalation), because the project leadership failed to identify sources for all of the finance needed to complete the project.

Identifying suitable sources of finance is the first step in planning finance for a project. Until sources for all of the finance have been identified, there is no point in spending a lot of time worrying about the cost of finance, the terms and conditions upon which it may be made available, or issues to do with evaluating risk.

DEBT, EQUITY AND AID

Finance for projects falls into two main categories:

1. **debt**, which has the essential characteristic that the borrower has the obligation to repay. Debt also usually carries an obligation to pay interest and to adhere to a prearranged repayment schedule. The lender has a priority claim if the borrower goes into liquidation.
2. **equity**, defined broadly as funds subscribed by shareholders in the project from their own resources. There is no guarantee that a dividend will be paid, so the investors may lose their money if the project fails to perform. Equity investors have the last claim if the project goes into liquidation.

Both debt and equity are available in a wide variety of forms distinguished in terms of, for example, the cost of the finance, the terms upon which it is available, limitations on its use for certain purposes or project activities, and the obligations of the respective parties to one another.

A special case of debt financing is *project financing*, defined as the financing of a project in which the lender relies for repayment primarily on the net cash flow generated by the project, and the main security is the assets of the project itself (see Section 10.5 below).

The cost and terms on which finance is made available always reflect the financiers' views about the riskiness of the project. However, raising equity is usually more difficult than raising debt finance. Once the available equity is committed to financing a particular element of a project, it is usually extremely difficult to obtain additional equity finance for other elements for which debt finance has not been foreseen, or not planned in advance, or for cost overruns.

The proportion of equity to debt in the financing of a major project has important consequences for the appraisal of the project. It also influences the way investors and lenders perceive the commitment of the project sponsor to the project. The debt/equity ratio would be expected to conform to certain accepted norms which depend on national, industrial and financial practice, and which tend to change with time, so that selecting the appropriate ratio for a particular project calls for detailed specialized knowledge of the investment market.

Aid is another kind of finance for projects. *Grant aid* is a direct gift of money normally made by one government (or a multinational supragovernment agency such as the World Bank) to another government. Grant aid is usually intended to help less developed countries meet social or community welfare objectives. It is rarely available for major industry projects (exceptions are, for example, hydroelectric power schemes intended to support rural electrification). Also, it is seldom entirely free from obligations on the recipient government and, although the strings may not be directly attached to the project, they may influence the way the project can be managed.

Other forms of aid, such as subsidized credit for exports, or aid-plus-credit packages (*'crédit-mixte'*) are best considered as forms of debt financing.

PROJECT ACTIVITIES AND THEIR RELATION TO SOURCES OF FINANCE

At an early stage of planning, project activities should be related to sources of finance. This will help to distinguish those activities which could be financed by equity from those which could be financed by debt. Because debt implies an obligation to repay, it is often not available for high risk activities such as minerals exploration, research or project feasibility work.

Equity, on the other hand, can usually be applied to all project related activities, provided the shareholders are prepared to bear the cost. In practice, however, difficulties may arise when, for example, the equity available for a project is denominated in a currency which cannot be converted to currencies which contractors or equipment suppliers are willing to accept.

As soon as possible in the planning of finance, the sources of debt and equity should be put into more specific categories. We need to know, for example, whether a specific source of debt (or equity) exists for a particular project expenditure, and whether, given the characteristics of the project we are considering, it can be made available when it is needed.

The main sources of equity finance are:

- corporate cash flow generated by existing business operations;
- corporate or individual investors, or funds raised through the stock markets;
- joint venture partners;
- government subscriptions;
- multilateral investment institutions such as the International Finance Corporation of the World Bank (IFC), and the investment arms of regional development banks.

The main sources of debt finance are:

- commercial banks;
- multilateral lending institutions such as the World Bank, and the European Investment Bank (EIB);
- suppliers of equipment and services for the project;
- suppliers of raw materials to the project;
- buyers of output from the project;
- aid institutions.

10.4 Raising finance

SOURCES AND METHODS

Finance for projects may be raised from:

- *Shareholders*, who may be public or private institutions or individuals, provide equity for use directly in the financing of certain project activities or as collateral for obtaining debt finance.
- *The stock market*. Shares in a company may be regarded as evidence of a commitment to the company's aims, or as title to ownership in the company. Commitment generally means at least the intention to support the company even in difficult times.

However, shares (titles to ownership) may be traded in the stock market. Shareholders who are primarily interested in this aspect of investment seek to maximize their returns (and the directors of companies may have a legal obligation to maximize shareholders' returns as dividends, and their assets). So shareholders are free to trade their shareholdings to the extent necessary to get the highest returns.

But the stock market does not price *future* income rationally. It places heavy emphasis on *present* rewards. Future rewards must be much greater than the rational level in order to persuade people to accept them. So the stock market undervalues returns which arise more than about 1 year in the future. In the UK it appears to undervalue returns 5 years in the future by about 40 per cent, and discounts cash flows more than 5 years in the future at about twice the rate of shorter-term cash flows. It follows that companies whose shares are valued by the stock market must adapt their business strategies accordingly. Thus, these companies are found to double the value of the dividends they pay out in order to keep shareholders who would otherwise take their money elsewhere. The money paid out in higher dividends is therefore lost for investment in projects other than those with very short payback. In other words, stock market activity of this kind inevitably leads to *short-termism* (see Appendix 2).

Banks and other financial institutions provide debt normally against security (for example, the borrower must provide collateral or convertible assets or, in the case of unsecured loans, must have a financial reputation for credit-worthiness acceptable to the lender).

Many banks offer debt financing. Keen competition between them helps to keep costs relatively low. Internationally active merchant banks often act as agents for arranging project finance by international investors. Access to international sources of finance can be advantageous in terms of, for example, lower costs, less government regulation, lower taxes, easier deposit requirements and reduced political risk.

Islamic banks provide finance according to the precepts of Shariy'ah law, which takes a significantly different view of the function of investment and profit than do Western commercial banks. Finance for projects in Islamic countries can be available to non-Islamic contractors. Negotiations can be expected to be complex and time consuming.

The main sources of international finance are the so-called Euromarkets (Eurocurrency, Eurobonds) for financial resources held outside their country of origin. Euromarkets offer lower interest rates as well as certain other advantages to large, highly credit-worthy corporate borrowers.

Export projects

Some special arrangements exist for financing export projects. Here, additional risks influence the choice of financing method. They include the following:

- the status of the project sponsor;
- the international market for projects of that type (and implications for the exporting country's balance of trade);
- the structure of the proposed export deal;
- the required length of credit and any periods of grace;
- the complexity that can be managed by the project sponsor (as well as by the exporter).

Many financial instruments have been developed to facilitate export trade. We will consider only a few from the project manager's point of view of ease of access and use, cost and risk sharing. In the following, the project sponsor (or owner) is referred to as the buyer, and the exporter – usually a contractor – as the seller.

Letters of credit, also called bills of exchange The buyer's banker issues a letter of credit as proof that the seller will be paid subject to specified terms. Letters of credit are negotiable, i.e. they can be traded (at a discount) for cash. They are relatively low risk instruments suitable for short duration deals involving relatively small sums, are easy to organize and cost relatively little.

Export credits are financed by banks and backed by insurance cover. The bank has no recourse to the exporter if the buyer defaults. In a 'supplier credit', the exporter – typically a contractor – obtains a loan from a bank and insurance cover from a specialized insurer. The terms of these arrangements are often subsidized by governments to promote export trade. The exporter should therefore be able to negotiate a competitive contract with the buyer. An additional advantage to the buyer is that he has only the exporter to deal with – the exporter makes all other arrangements. A disadvantage is that the supply contract is subject to approval by the seller's government (*see* Fig. 10.1).

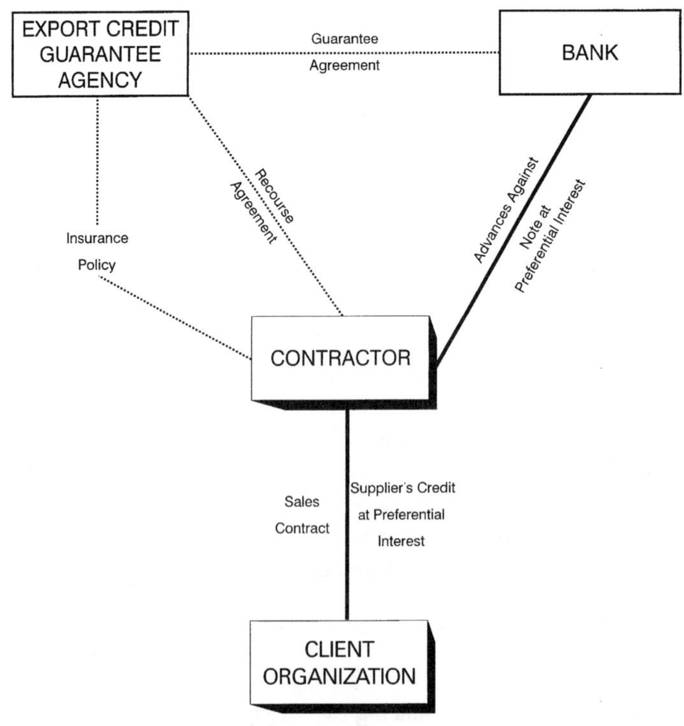

Fig. 10.1 Supplier credit.

However, contractors are rarely able to hold large amounts of debt for long. Consequently, major projects are typically financed by 'buyer's credit', where the contractor is relieved of the necessity to hold debt.

In 'buyer's credit', the buyer, usually a government or a parastatal agency, borrows direct from a lender in the seller's country. The seller is paid by the lending institution against evidence of satisfactory supply. Buyer's credit often falls within a government-to-government line of credit, which enables the buyer to negotiate contracts for various goods or services in the lending country. However, the buyer now has two channels of negotiation to manage: with the seller and with the lending institution in the seller's country (*see* Fig. 10.2).

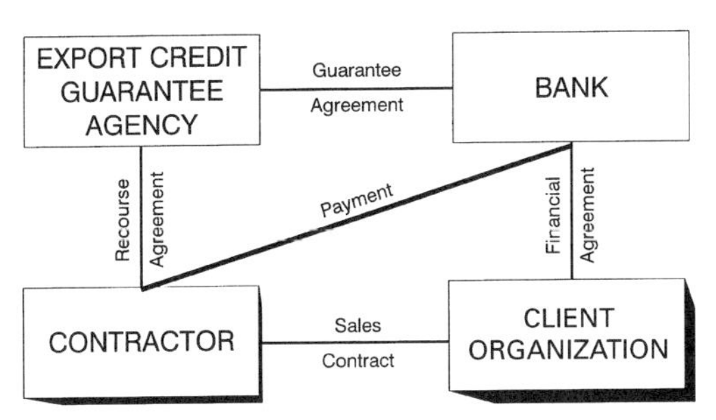

Fig. 10.2 Buyer credit.

Export credits are usually conditional on the greater part of the supply being obtained from the seller's country. The proportions which are admissible vary from country to country, and are subject to international agreements which often seem in practice to be more honoured in the breach than in the spirit. In general, however, export credits are a popular way of financing both large and small export projects, provided care is taken with the terms of credit insurance.

Aid Export credits are sometimes combined with forms of aid (as in '*crédit mixte*' or the UK 'aid and trade' provision). Such arrangements are suitable for large prestigious projects where government support needs to be mobilized to establish a position against international competition. While the eventual terms may be exceptionally favourable, the project promoters will have to commit a lot of senior management effort to satisfying bureaucratic requirements.

This is true of mobilizing aid finance in general. There are basically two forms of aid: *project aid* for specific projects, highly structured and formalized in accordance with the donor institutions' project appraisal procedures; and *programme aid*, which is intended to finance imports (including projects, though usually small ones) in return for sectoral policy reform. Programme aid leads to shopping lists of imports which may be specific, allowing individual projects to be identified, or may include only general aims.

Aid finance is normally government-to-government (bilateral), when it is usually, but not always, tied to supplies from the donor country, or is from a multilateral institution, when use is not tied to particular sources of supply. There is a tendency for the multilaterals to blur the distinction between development aid and commercial lending through low-interest (soft) loans and grant aid used to dilute interest payments, or by co-financing, which involves the participation of other donor or lending agencies working with the multilateral institutions.

This multiplicity of options and sources serves to increase the amount of management effort needed to secure the financing of a project which involves an aid component. Nevertheless, aid provides a useful niche market for financing some projects, especially those where open competitive bidding is deemed to be inappropriate. Examples include those where only one seller is considered to be capable of carrying out the project; certain high risk projects; and specialized areas such as maintenance, emergency or training projects.

JUSTIFYING THE PROJECT

Because there are many possible sources of debt and equity, it is usually possible to identify, at least in principle, a mix of potential sources for all the various financial components of a project. The real problem is to secure the financing of the project on terms which promote the project's chance of success.

As a first step, recall the purpose of the financing: it is simply to improve the project's contribution to the owner's financial performance. Consider the examples in Table 10.2(a) and (b). Clearly, the higher the debt:equity ratio, the more impressive will be the project's financial performance in relation to the equity employed. From the point of view of the lending banks, a 'good quality' debt:equity ratio would be about 2:1. Typical ratios may, however, go to 3:1 (i.e. 75 per cent of the project investment is debt). But 'financial engineering' by increasing the debt:equity ratio so as to improve returns to shareholders strengthens the illusion that average financial returns can be much higher than, and in some way divorced from, growth of the economy. Companies then tend to get out of what they perceive as low growth businesses faster than they invest in new (higher growth) projects. (Of course, the financial performance of a project will also depend on the taxation policies under which it operates. Consideration of this is outside our present scope.)

Table 10.2 The influence of debt on financial performance

(a) No debt: IRR = 21.0%

	Year					
	0	1	2	3	4	Total
Capex	(100)					(100)
Net revenue	0	20	40	50	60	170
Cash flow	(100)	20	40	50	60	70

(b) 60% debt and 40% equity: IRR = 31.2%

	Year					
	0	1	2	3	4	Total
Capex	(100)					(100)
Net revenue	0	20	40	50	60	170
Loan	60					60
Repayments	0	(15)	(15)	(15)	(15)	(60)
Interest @ 10%	(6)	(4.5)	(3)	(1.5)	(15)	
Cash flow	(40)	(1)	20.5	32	43.5	55

The project owner's aims in raising finance may be summarized as:

- to raise the necessary amount of money at the times and in the currencies required by the project;
- to minimize costs;
- to minimize risks by sharing them with other parties in the project;

- to maintain flexibility and control, including the possibility of rescheduling if necessary;
- to be able to pay dividends to shareholders.

The lender's objectives are rather different:

- to be satisfied that debts (principal, interest and fees) will be repaid on time;
- to be satisfied that there is adequate security and recourse in the event that the borrower defaults on payment;
- for shareholders, that satisfactory dividends will be paid.

These differences in viewpoint are least when the project is financed from corporate resources. If external financing is needed in addition, the lender (or investor) will take a view that the corporation is backing the project, so that the main risk is of the corporation defaulting. For financing raised by established corporations, this is normally the minimum cost scenario as well.

10.5 Project finance

INTRODUCTION

Project finance is finance provided against the assets of, and rights in, a particular project rather than against the borrower's balance sheet. This means that the lender has, in principle, no recourse to the borrower's balance sheet. The lender therefore takes the risk that the project will fail to come up to its sponsors' expectations. Project finance is advantageous in a number of situations, e.g.:

- when a government decides to privatize significant parts of the public sector;
- when a government decides to try to increase foreign capital investment in the country at no cost to the country;
- for governments, to share the cost and risk of exploiting natural resources;
- for multinational corporations, to protect their corporate balance sheets from the impact of large projects;
- for companies in general, to avoid constraints on corporate borrowing;
- for project sponsors, where one party is not able by itself to take on a major project, to spread the risk among several parties so as to lessen the financial impact and to increase their capacity to undertake more projects.

These advantages must be paid for in the higher cost of project financing to compensate the lending banks for the greater risk which they accept, and in the extensive legal and *quasi*-legal requirements as to what must be done and what must not be done, to satisfy the lender that the project is *bankable*.

These requirements are fundamentally doctrinal (in that they relate to the lender's condition of confidence and belief that the project will come up to expectations). Their influence on the management of project risk is therefore categorically different from the *physical* risks with which project managers are perhaps more familiar. So it is all the more important that project leaders and managers appreciate the position of the lender (a bank or a consortium of banks) and the lender's legal advisers regarding project risk.

From their perspective the main issues are:

- the lender expects the loan (principal plus interest) to be repaid from the future cash flows of the project;
- the lender considers the assets of the project (including any rights held by the project) as security for the loan;

- the lender does not expect any third party guarantee to secure the loan directly;
- the lender does expect support and undertakings to cover some risks to be provided by third parties with an interest in the project;
- the lender, while accepting to take some risk in the project, will avoid taking substantial risk, and will always look at alternative ways of securing repayment;
- the lender sees himself as a facilitator, not as a risk-taking investor. As facilitator, the lender simply provides funds to cover the shortfall between the total project cost and what is available from the sponsors and other interested parties.

NON-RECOURSE AND LIMITED RECOURSE FINANCING

What is described above is essentially *non-recourse* financing. 'Non-recourse' means that in case of default by the borrower, the lender's only remedy is to enforce his security on the project's assets and the revenue which the project assets generate. The lender has no right to act against the borrower himself (or any associated parties), although the borrower will remain liable for his warranties and covenants.

The lender therefore normally requires the borrower to covenant to repay. This avoids any doubt about the enforceability of security, and gives the lender access to the borrower's non-project assets. It also strengthens the lender's position relative to the project sponsors. For these reasons, project sponsors often create a 'single purpose vehicle' (i.e. the *project company*) to be the borrower and carry on the project, but which has no other assets. This separates liability for the loan from the project sponsors. Consequently, lenders usually try to obtain some commitment to the project from the sponsors; hence 'non-recourse' becomes in practice 'limited recourse' financing.

SECURITY

The type of security in project financing is an important factor, because non-project assets are in principle excluded from the security arrangements. What is possible depends on the relevant jurisdiction. The main types of security are:

- **Mortgage.** This is the normal method of taking security over land, fixed assets and machinery. The precise form of security depends on local laws and the remedies available for enforcement in case of default. Cross-border financing raises special problems with mortgage-based security.
- **Blocked accounts.** Special accounts may be set up to receive project revenues in order to provide security to lenders. The terms under which the project company may operate the accounts must be agreed between the project company, the lender and the bank at which the special account is held.
- **Assignment of contract benefits.** An example of this is the assignment of benefits under off-take or supply contracts. Assignments are difficult to negotiate in view of the lender's aim to perfect its security and ensure ease of enforcement.
- **Shares.** Lenders may require security over the holdings of equity participants in the project company together with a ban on the transfer of these shares without the lender's consent, in addition to other securities. The mechanics of this form of security depend on the country in which the project company is incorporated.
- **Completion guarantee.** When provided by the project sponsor, this is an undertaking by the sponsor to complete the project. As such, it does not usually give rise to a claim for any sum

of money. It may give rise to a claim for damages, but a court ruling on the matter may calculate damages at an amount likely to be different from the amount of the lender's loan. Lenders may therefore seek a guarantee in the form of a letter of credit or a performance bond from a bank or an insurance company.

In developing the overall security arrangements, the lender will consider several additional factors. These include *consents* required from government authorities; the appointment of an *agent* to hold the securities and act on behalf of the lenders; *registration* requirements to ensure the validity, priorities and enforceability of the securities; the *proper law* governing project transactions so as to make sure of the effectiveness of the security arrangements in the country of the project.

10.6 Lending banks' risk assessment

Many project risks are interrelated and interact, making it difficult to decide where they originate and to which party they should be allocated (on the basic principle that responsibility for carrying the risk should be allocated to the party best able to manage it). However, it is convenient to classify risks into broad categories, which helps understanding of potential interactions and impacts. For lenders, a typical classification is:

- pre-completion risks
- post-completion risks
- financial risks
- political risks.

PRE-COMPLETION RISKS

The lender often retains independent consultants to advise on feasibility studies prepared before award of any design or construction contracts, and on planning of the construction phase, because this determines the project completion date and therefore the time when revenues begin to flow. The consultants' report on the project, its objectives, viability, structure, strengths and weaknesses, the parties to it and the relationships between them is the primary basis for the lender's assessment of the *bankability* of the project.

The lender's main concerns will be with:

- sponsors and investors, their relationships and project responsibilities, especially the likely consequences of disagreement between them.
- the technical competence of contractors, their track-record, and the proposed contractual arrangements. Lenders may recommend that the project sponsors prefer contractors' bids which offer the best security, rather than the lowest price.
- technical risks – lenders normally prefer projects with well-proven technology, where the contractor has relevant experience of the design and construction of similar projects.
- environmental risks, which are potentially enormously damaging. Lenders will be keenly interested in environmental impact studies, and – especially in countries where customary law applies – resolution of site land ownership rights.

POST-COMPLETION RISKS

Remembering that the prime concern of the lender is repayment of the project loan plus interest, the lender can be expected to view project risks differently from the equity shareholders. The latter take a generally longer-term view and, as the main risk takers, expect a higher return. But because the project is justified essentially in terms of its revenue-generating capability, security for *continued* revenue generation is crucial.

Lenders will look to the borrower to provide evidence for:

- proper provisions for operation and maintenance, such as ongoing technical support from contractors and equipment vendors;
- raw materials supply, especially security of supply and stable prices;
- product sales, e.g. long-term contracts with one or more off-take partners. These can sometimes cause severe problems, e.g. the 'take or pay' contracts entered into by British Gas for North Sea gas, widely reported in the press during 1995.

FINANCIAL RISKS

All project risks are eventually expressed in terms of financial risk. The main sources of information for study are the project's cash flow statement and pro-forma balance sheets, profit and loss statements, and statements as to the source and application of funds. These instruments, which demand great effort in preparation, should be quite comprehensive and should reflect actual business as accurately as possible. They show where revenue is expected to arise, and where it is expected to increase or decrease in the business cycle. They show routine maintenance shut-downs, payment of taxes, etc.

Project financiers like to say, 'Cash is sanity, profit is vanity'. The lender's main concern is the project's *cash* position. If there is not a sufficient flow of revenue in the form of cash, the project is dead.

Sensitivity analysis should show how robust the project is to variations in costs and income. The lender will be interested mainly in 'worst case' scenarios, because he needs to know to what extent project economics can deteriorate before the project fails to meet its debt servicing obligations.

In analysing future cash flow projections the amount of debt that can be serviced throughout the life of the project will be established. The loan agreement will require the project company to maintain a minimum cover ratio to provide for repayments of principal and interest. A typical ratio is 1.5 times the net cash flow to meet the next principal and interest payments due. Cover can be improved by deposits into an escrow account to secure future payments. The lender will consider arrangements like these to contribute to the security of the project loan.

Other financial risks which the borrower will be expected to cover include:

- cost overruns
- foreign exchange risks
- interest rate movements
- inflation.

POLITICAL RISKS

A project will normally be considered for project financing only if the sponsors can show that the political climate in the country where the project is to be located is stable, and that the local legislative framework of regulation and recourse is appropriately structured. It is difficult to establish non- or limited-recourse financing in a country which has a history of political upheaval.

Force majeure

Force majeure, being in broad terms situations over which none of the parties to an agreement has control, can easily be drafted as a catch-all for risks that no-one wants to be responsible for. From the lender's point of view, a risk shared is a potential problem doubled. The lender wants to know who is responsible for clearing up a problem should one occur, and does not want to get into an argument over who is responsible for what. Therefore it is in the interest of the lender for the definition of force majeure risk to be specified as narrowly as possible.

Lenders may seek to reduce political force majeure by, for example,:

- requiring that the project's income arises from foreign buyers and is paid into an overseas account. This protects the project cash flow from acts of the project host country government, but is not possible if the only product buyer is a state agency.
- obtaining formal assurances from the host government in relation to areas such as taxation, remittance of revenues and profits overseas, expropriation or nationalization of project assets, import duties on project-related supplies, royalties, enforcement of lender's security, etc. However, even if obtainable, none of these undertakings can be enforced.

Risk allocation

The lender seeks to accept only such risks as he can control. The lender will therefore consider the project as a whole to see where risks have been identified, to whom they have been allocated and to what extent he can draw comfort from these risk allocations.

A fixed price contract with suitable liquidated damages for delay or defective specifications is the main instrument for allocating completion risk to the contractor and manufacturers of project equipment. A fixed price contract may also be the means of transferring some of the force majeure risk to the contractor, or the risk of regulatory changes requiring redesign, thus transferring at least some risk away from the project company and the project revenues assigned to debt repayment.

Protection can be enhanced by calling for performance and completion bonds from the contractor, which lay off the risk further, as a credit issue to a bank guarantee for payment. Because only large, commercially strong contractors can provide the bonds required, this mechanism provides additional comfort to lenders.

10.7 Seeking financial proposals

Identifying the potential mix of financing sources and justifying the need do not solve the problem of financing the project. We now have to convert the potential mix into commitment, or at least into proposals capable of being accepted. This step is especially important if there are difficulties about the location of the project or the status of the project sponsor. First of all, we have to distinguish between sources of finance which can only be approached by the project sponsor (or in some cases, by a government authority acting on behalf of the project sponsor), and those which may be approached by parties other than the project sponsor, such as contractors or equipment suppliers.

The World Bank and regional development banks can be approached only by governments or by project sponsors who are acting as the agent of a government body. Export credit is only available in support of exports from the country of the suppliers, and is normally obtained through the suppliers or contractors concerned. Commercial bank finance can be accessed by project sponsors, or by suppliers or contractors in support of their bids. Aid is arranged through a combination of project sponsor, suppliers and their respective governments.

The project sponsor's ability to get good terms for finance is considerably enhanced if he has a 'borrowing base' such as a portfolio of projects. If the portfolio is well planned, with a good balance of reliable projects as well as high risk ones, lenders may be inclined to accept the portfolio as a risk dilution device. This will reduce the cost of borrowing to the project sponsor.

If a project sponsor intends to ask suppliers or contractors for supporting finance, it is crucially important for the enquiry documents to specify this and the nature of the support requested. The relevant sections of the enquiry documents are best prepared by specialists to ensure that the project sponsor's freedom of action is maximized, and at the same time to motivate the support of the suppliers or contractors and their respective financial institutions.

The project sponsor will do well to cultivate contacts with several potential sources of finance before making a formal enquiry. Otherwise, the project sponsor has no alternative but to 'cold call'. Because financing depends primarily on confidence, this is the approach least likely to be successful.

Once contact is established, the finalization of financial arrangements is a matter for negotiation. Each project is unique; the terms for its financing will also be special, though based on precedent to the extent that the lender's experience allows. We have discussed this general background above, and it should help the project sponsor to orient his approach. It is perhaps useful to remember Keynes' dictum that 'if you owe a bank £100, you have a problem; if you owe it £1 million, it has a problem'. Allow two orders of magnitude for project borrowing today.

One final point about approaches to financial institutions. Money lending, like all other trades based on art rather than science, relies on jargon to mystify and impress its clients. Be sure you know you know precisely what is meant by bankers' jargon; many words have technical meanings different from their meanings in ordinary usage. If in doubt, ask for clarification.

10.8 Unconventional financing

A number of unconventional methods of raising finance for projects have evolved in order to try to overcome the disadvantages of the conventional markets discussed above. Among the aims of these methods are:

- simplicity;
- to generate convertible currency;
- to gain access to markets which are otherwise restricted;
- to avoid a drain on balance of payments;
- to accelerate development of technical skills;
- to comply with political objectives;
- above all, to reduce or shift the balance of risk.

Like other attempts to alleviate risk, they incur costs. It is a matter of nice judgement whether they actually achieve their aims.

LEASING

This makes available the use of project assets through off-balance sheet financing, and may attract favourable tax allowances. However, tax laws are subject to change at short notice under the influence of arbitrary changes in government policy, so the apparent attraction of leasing may turn out to be illusory.

FORFAITING

This makes available finance through the sale of financial instruments due to mature at some time in the future. The cost in terms of fees and discounting can be very high.

COUNTER-TRADE

In counter-trade, the seller accepts (or makes arrangements for a third party to accept) goods or services in lieu of cash payments. Counter-trade may take the form of a simple exchange or 'barter', or the project may be paid for with its own output, as in 'buy-back'. There are many ways of effecting counter-trade, all expensive, cumbersome and tricky to negotiate. For these reasons, counter-trade is unloved by both project contractors and the financial community. There are also philosophical issues to do with interference with free trade which are generally believed to be undesirable aspects of counter-trade. On the other hand, buy-back is said by its proponents, usually politicians in developing countries, to be a satisfactory means of encouraging technology suppliers to commit themselves to the long-term transfer of their technology in complex projects (Fig. 10.3).

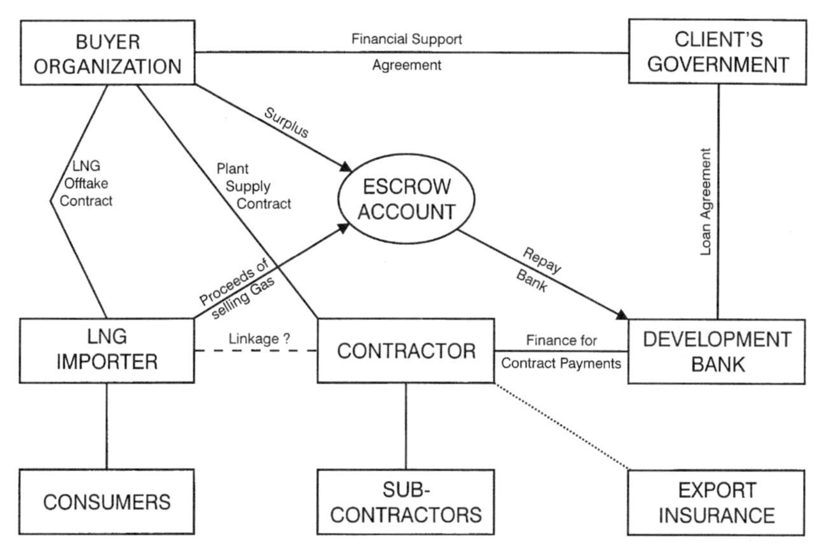

Fig. 10.3 Typical 'buy-back' for an LNG project. There is often a linkage between the LNG importer and the contractor. For example, the contractor is a subsidiary of the importer organization.

SWITCH TRADING

A technique for making use, *via* a third party, of uncleared credit surpluses arising from bilateral trade agreements. For example, if country B has a credit surplus with country C, exports from A to B may be financed by payments effected through a switch trader from C to A. Needless to say, this can lead to some very complex and expensive arrangements.

OFFSET

This requires the exporter of a technically advanced project to incorporate an agreed value of materials, equipment and/or services supplied by the buyer. Offset may be extended to require the exporter, as owner of the technology, to set up local production facilities associated with the project. The idea is again to provide evidence of the seller's long-term commitment to the development aims of the buyer.

FRANCHISE FINANCING

This requires engineering and construction contractors to become equity-holding joint venture partners in the projects they design and build, and thus to remain associated with implementation and operation of the project over a number of years. A hoped-for consequence is that this will result in better quality projects, because the contractors and/or licensors will share responsibility for efficient long-term management.

DEBT/EQUITY SWAPPING

This is a device to encourage investment by technology owners. Typically, a multinational technology owner buys host country debt at a discount. The debt is redeemed in local currency at a favourable rate of exchange for the purpose of setting up a local company, which may then use the transferred technology to generate foreign exchange, replace imports and create local employment.

BUILD–OPERATE–TRANSFER (BOT)

In a typical BOT project, a government grants a concession to a project company to build a facility and operate it on a commercial basis. The project company finances design and construction of the facility with money from equity subscribers and/or lending institutions. Loans are repaid from tariffs paid by the government under an off-take agreement during the life of the concession. At the end of the concession, when (hopefully) the project company has earned a return for its shareholders, the facility is transferred to the government (*see* Fig. 10.4).

BOT allows large, economically fundamental, projects (energy projects are a good example) to be developed without governments having to commit their possibly limited financial resources. The BOT concept is therefore becoming increasingly popular with the governments of developing countries as a means of integrating major project development, technology transfer and the trend towards privatization. However, the contractual arrangements in a BOT project are extremely complex, and it is frequently found that the legislative framework existing in developing countries, or in countries in transition from a centrally planned to a market oriented economy, is inadequate for BOT contracts. In addition, BOT calls for exceptionally long-term commitments by the contracting parties and is vulnerable to currency fluctuation and to political risk. These factors tend to constrain exploitation of the BOT approach.

10.9 Financing problems and their implications for the management of projects

Although, as we have seen, there are so many and varied sources of finance for projects that it is rarely impossible to find potential lenders or subscribers, establishing a workable financial package is difficult and fraught with pitfalls for the actual implementation of the project. Project managers are wise to take note of these, for the pitfalls snare their feet more often than they trip up corporate money men.

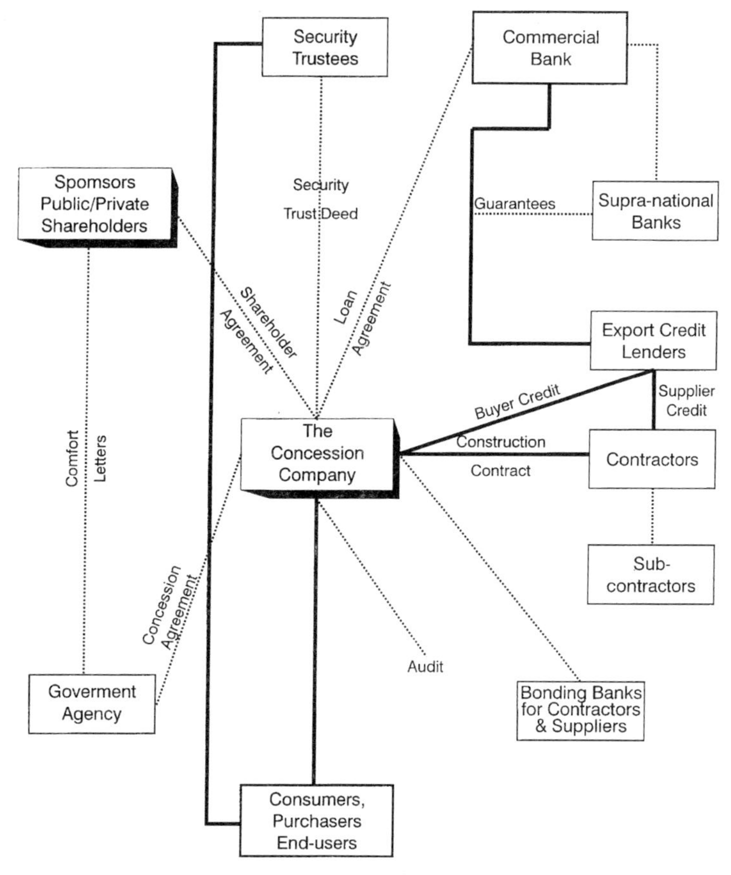

Fig. 10.4 Financing structure for a BOT project.

DEBT SERVICING

This is the repayment of the sum borrowed (that is, the 'principal') plus interest repayments. In the final feasibility study, project cash flows will be shown to accommodate debt servicing requirements comfortably. If, eventually, they do not, then the project and the project manager are in serious trouble, for the cost of financing unforeseen adverse cash flows is very high. Unexpected adverse cash flows are most likely to arise during the final stages of construction and completion of project implementation. This is often the time when debt servicing payments fall due, while the project is not yet generating income.

COSTS OF FINANCING

Fees are payable to the financial institutions which organize the money for the project. Often, more than one institution is involved, for example, as agent for the borrower or as actual providers of finance. Fees are payable for a wide range of services, which may not be immediately apparent when the project's financial planning is being considered. The main costs of financing a project are (typical rates at the time of writing):

- bank's arrangement fee – 0.5 per cent;
- bank's management fee – 0.25 per cent per year;

- interest rate – typically fixed in the loan agreement for a stated period beginning on the date the loan agreement is signed; the rate is also dependent on the strength of the currency in which the loan is denominated;
- insurance – depends on the perceived risk and is based on the amount insured, typically 85 per cent of the project contract value;
- standby fee – negotiable.

If the project go-ahead is deferred, standby fees are payable to the financial institutions involved in keeping the finance available. Standby fees will almost certainly not have been foreseen in the original project plan.

INSUFFICIENT FUNDS

Despite the obvious problems, it is very tempting for enthusiastic project promoters to get the project started in anticipation of a final tranche of finance being available at some later date. It may not be, or it may be available only on extravagant terms. Frequent causes of insufficient finance include :

- miscalculation of the actual capital costs required;
- poor definition of actual work scope, which leads to more work being necessary than was envisaged when the finance was sought and obtained;
- changes to project work scope accepted without considering their cost implications (a frequent problem arises from a decision to fast-track the project in the belief that this will have little or no effect on costs);
- underestimating the cost of working capital and/or spares;
- project delay – even when finance for the total cost of the project is available, if money is not available to cover project expenditures at the required time, the project will suffer from insufficient funding.

CURRENCY

Project costs are denominated in one or more specific currencies. If financial input and output are denominated in different currencies, exchange rate fluctuations may cause difficulty. For example, petroleum project revenues are usually denominated in the US dollar. Project finance in a currency which tends to appreciate against the US dollar will embody an exchange risk which may be extremely costly for project implementation.

Most exchange rate risks can be hedged by, for example, using currency forward markets, interest rate swaps, options and other instruments. Hedge mechanisms are themselves risky, especially in the hands of amateur money managers. They are also expensive. Project managers should try to ensure that project input and output are, so far as is possible, denominated and balanced in the same currencies.

Inflation raises similar issues if not adequately accounted for in project planning. Note in particular contractors' 'overnight' cost estimates. 'Overnight' here means that inflation and interest payable during the period of design and construction are excluded from the estimated cost.

SECURITY

Security is normally required, in some form, for debt finance. In addition to conventional types of collateral, there is increasing use of project assets, rights to project resources (e.g. oil or gas in the ground) and rights to project outputs as security. Another form of security much used in industrialization projects in developing countries is the sovereign guarantee of the government of the country where the project is located.

In recent years, in some parts of the world, sovereign guarantees have lost some of their virtue because of debt crises which colour the judgement of some finance ministries. Financial institutions are then moved to pay more attention to the assets represented by the project and its value as a means of exploiting natural resources or developing markets. Although this is sometimes considered to be a novel development, it is of course the way in which, historically, most large projects were financed, a fact possibly obscured by the increasing sophistication of both projects and the financial markets.

INVESTMENT RISK

The providers of finance to a project are in fact committing their resources to an investment in the country where the project is to be located. They must therefore take a view of *country* risk as well as project risk. A view of country risk is provided by the country credit ratings made by official credit insurance agencies (e.g. ECGD in the UK). These ratings should be considered together with the limitations on credit availability and the interest and other terms imposed by the lending government. Credit ratings reflect official perceptions of the riskiness of investment in the recipient country and may of course be coloured by political objectives.

Institutional perceptions are reflected in country risk investment ratings made on the basis of surveys of opinion among bankers, brokers and analysts. These surveys take account of factors such as the country's overall economic performance, political risk (e.g. of expropriation, of government interference with exchange rates, repatriation of profits, etc.), debt indicators, debt default or rescheduling, credit rating, access to various sources of finance and prevailing credit terms. Country risk surveys do not, of course, reflect the credit rating of a particular project sponsor organization. Moreover, investors' perceptions are likely to be strongly influenced by media reports of economic activity in potential investment locations.

Political uncertainty makes for an unattractive investment climate. Political risk can be lessened by arranging several agreements between the project owner and the various government bodies involved with the project (but note that competition among departments of government is a fact of bureaucratic life everywhere), and by arranging finance through lenders in several countries.

Political risks can be insured against, although this is expensive. The World Bank has set up the 'multilateral investment guarantee agency' (MIGA), the purpose of which is to provide insurance to companies investing in the Third World against the expropriation of their assets, and to cover certain other political risks.

INSURANCE

Although insurance matters are normally handled by a specialized corporate department, it is the project manager's responsibility to ensure that his project is adequately covered. Responsibility for insurance, including liabilities and indemnities, must be defined at the ITB stage of contract negotiation. Contractors must be required to insure their own staff and equipment, and to cover general third party liabilities up to a specified limit (typically at least US$10 million).

Corporate guidance should be obtained on statutory and local insurance requirements, and for other insurances normally needed or to be considered, e.g.:

- general Third Party liability
- construction all risks
- fire, lightning and explosion
- transportation of equipment and materials to the operational site
- general plant material and equipment damage
- motor vehicles
- staff (possibly including fidelity guarantees).

Offshore projects involve special insurance considerations, as do exploration and production projects (notably in connection with blow-outs and pollution). In some locations, earthquake and unusual weather conditions insurance may be appropriate.

It is an unwise project manager who assumes that someone else is liable and that this someone else has taken out insurance. Make sure contracts stipulate who is liable and who must take out insurance, and then check that the insurance has indeed been effected.

10.10 Summary

The characteristic features of projects impose special requirements for financing project capital. While project managers are seldom responsible for making the financial arrangements, these arrangements can influence the manageability of the project, and their implications should be understood by project managers and project leaders alike. Financing strategy is a crucial component of project development strategy. Financing adds to the complexity of the project. It demands early consideration and management commitment if a satisfactory outcome is to be achieved.

Many different types of finance, from many sources, are potentially available. The suitability of a particular type is coloured by the financier's perception of project risk, which may well be different from that of the project sponsor. Usually, a mix of various types of finance proves most suitable, although the terms upon which each type of finance is made available is always a matter of negotiation with the various sources.

The response of potential sources of finance is essentially the response of a market. Sources (or markets) respond to different kinds of project in terms of relative ease of access to finance, its use, cost and risk sharing largely as a function of the project type, scale and whether it is an export project or not. In addition to conventional methods, a rather wide range of unconventional methods of financing has evolved, mainly intended to shift the balance of risk, but also with a view to improving access to finance in otherwise difficult situations.

Accessing finance inevitably involves justification of project activities. Financiers expect project sponsors to justify their projects in terms that are understandable and acceptable to financiers. Usually, this requires extensive information about how the project will be managed, as well as estimates and forecasts of its expected performance. The approach and subsequent detailed negotiations will be time-consuming – another reason for early and careful preparation on the part of the project leader.

At the same time, indeed as part of their preparation, project leaders and managers must be aware of, and on the alert for, problems in financing and their implications for the manageability of projects, because insufficient or untimely availability of finance for any phase of the project, or situations that increase project risk, will inevitably reduce the chances of project success.

11
Licensing Technology

11.1 Introduction

Most projects involve bringing 'knowhow' from outside the project team. Sometimes all the required knowhow can be obtained from in-company sources. More often, it has to be obtained from other sources. In any case, a contract (or contract-like) relationship will arise between the project and the source(s) of knowhow. There are three main sources of knowhow:

- **licensors** – organizations, or occasionally individuals, whose business is, in the present context, to sell you rights to practise technology which the licensor has developed and practised himself, or acquired somehow from elsewhere;
- **consultants** – organizations or individuals whose business is essentially to sell 'expert' advice;
- **contractors** – organizations (rarely individuals) whose business is essentially to sell a package of goods and services which they have put together for a particular purpose such as a project.

These classes often overlap. Licensors and contractors sometimes act as consultants. Consultants often act as licensors or contractors. So we should regard the classification above simply as a convenience for our present purposes. But the classification carries implications which should be taken into account in setting up working relationships. For example, a company appointed as consultants to advise on the feasibility of a project should not normally be invited to bid for work as the contractor to design and construct the project.

11.2 Licensors

WHAT IS A TECHNOLOGY LICENCE?

A licence is an agreement covering the supply of technical information from one party (the licensor, who has it) to another (the licensee, who wants to have and use it). A separate aspect is that of permission granted by a government to carry out specified activities reserved under law, and designed to stimulate technical or economic development. These two aspects of licensing are often closely linked, but here we are concerned with the first.

Licensing is one way to acquire technology. Other ways, such as developing it oneself, outright purchase or forming a joint venture, are more advantageous in some circumstances, less in others.

Licensing is also a way to exploit technology. There are, of course, other ways, and the pros and cons of these also largely depend on where you stand, notably on whether you have the technology and want to exploit it by licensing out to others, or whether you need the technology and aim to acquire it by licensing in.

LICENSING OUT

Having spent a large amount of money and a lot of time developing a technology, why should any company want to license it, thereby risking the creation of more competitors? Possible reasons are:

- The technology was developed for internal use, but now needs larger scale operation than the licensor is prepared to fund.
- The technology has other applications than those for which it was originally developed, and the licensor seeks cooperation with companies active in these other areas.
- The licensor no longer wishes to exploit the technology himself.
- The technology is only partly developed, and further development requires a partner.
- Overseas markets cannot be penetrated because the technology requires technical marketing support beyond the capability of a sales agent.
- There are local restrictions on imports and it seems preferable to license local manufacture rather than set up a local subsidiary.
- Licensing may help to standardize industry practice, to the comparative advantage of the licensor.
- It may be more profitable in some market situations to collect royalties than to invest in a direct market position.
- Licensing, including to competitors, may cause the total market to grow in such a way that the licensor's share is worth more than it would be without licensing.

However, in addition to motives like these, would-be licensors need other attributes to make them attractive to licensees. They should:

- possess the required technology at the level required by the licensee;
- be prepared to license it;
- be likely to remain technologically competitive through continued R&D, the results of which will be shared with the licensee;
- understand the licensee's objectives and appreciate his technical and marketing strengths and weaknesses.

The list of attributes is now beginning to take on something of the appearance of an advertisement in a 'lonely hearts' magazine. What about the other party? If the sought-for relationship is to reach the desired consummation, what should one look for in a licensee?

LICENSING IN

The motives for licensing in are usually to do with technological or marketing diversification for either survival or expansion. The best candidates, from the licensor's point of view, are those who have some marketing expertise in areas where the licensor has none, and who are hungry for new technology – hence the considerable interest of licensing into the former CMEA (Comecon) countries and the more developed of the newly industrializing countries (nics).

However, enterprising companies which appreciate their need for technology at a genuinely appropriate level – which I propose to define as 'the most advanced which can be maintained in normal operation by normal means at the licensee site' – are by no means common – hence, most of the interest and a lot of the pain in licensing.

11.3 Choosing technology

Assume that we intend to choose a technology because:

1. we want to license it for use in our own plant; or
2. we want to license it to sell on to somebody else.

In either case, we are interested in being able to show (to our board, to our customers) that we have chosen the 'best' technology. The first question is, therefore: how do we choose what is demonstrably the best? There are a number of stages, depending on how far ahead of the field we aim to be.

TECHNOLOGICAL FORECASTING

This is a more or less black art which aims to work out the nature and efficacy of expected technological advances in a given field. Probably the most popular method of technological forecasting is the DELPHI interviewing technique, in which (essentially) you ask technical experts for their opinion. But because the most readily available technical experts are either already in-house or otherwise related to the organization asking the questions, the method tends to be incestuous unless special efforts are made to get outside contributions (which can be expensive), and there is always a problem with weighting the responses received. But it does help to develop management thinking about new technology needs and about the allocation of funds, specialist personnel, etc.

Technology assessment looks at the probable secondary consequences of hypothetical technologies (e.g. using morphological analysis) and tries to make explicit comparisons of alternatives as a basis for planning the orientation of development work.

Experience curves attempt to establish economic trends of technology development. Experience curves originated from the ideas of the Boston Consulting Group in 1968. They show that constant money cost per unit of added value declines by a characteristic amount each time accumulated experience doubles. So a log/log plot of cost (in constant money terms) against cumulative production approximates to a straight line with a slope characteristic of the production technology.

In practice, however, experience curves are more complicated than this. Consequently, they are reliable only for high growth products, because the characteristic cost decline is the result of unconstrained growth in product sales, which motivates producers to seek, for example, economies of scale or process innovation. This in turn stimulates the competition, which drives costs and prices down.

But the advent of radical new technologies changes the slope of the experience curve. A sharp increase in the cost of raw materials will distort the experience curve (Taylor and Craven, 1979) (*see* Fig. 11.1).

Despite these problems, experience curves can be useful for:

- forecasting price and cost trends when planning strategy, especially involving the pricing of new products (or existing products from new processes);
- monitoring the performance of existing products so as to set targets for new business (often involving the acquisition of new technology).

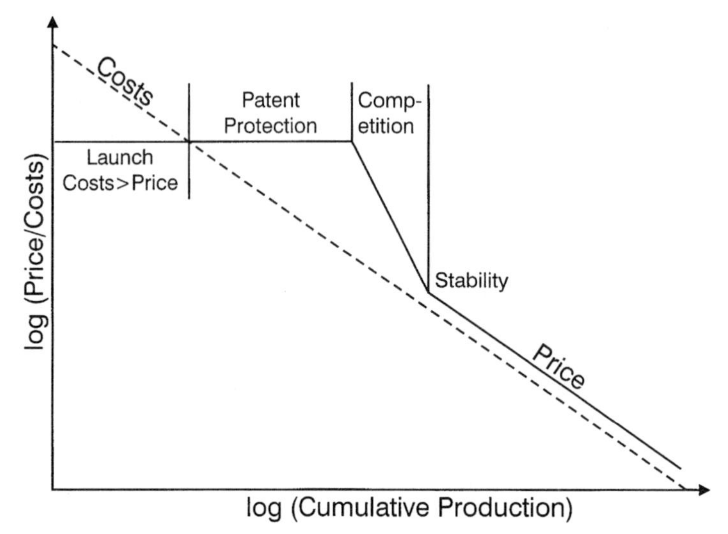

Fig. 11.1 The influence of market forces on the 'experience curve'.

The problems are basically those of interpretation and application, especially when price/cost trends are affected by external events, such as:

- fluctuations in the cost of energy;
- cost of meeting new safety or environmental legislation;
- government interference in pricing or market regulation.

The introduction of new technology often imposes higher costs on the innovator than the current industry-characteristic level, so that an act of faith by innovators or their backers is almost always called for. This is the reason behind the widespread feeling among businessmen that it is usually better to be second in a new technology than to be the front runner.

PROCESS SELECTION

Forecasting technology trends usually gives only a 'ball park' view of the technologies likely to be of interest for licensing. For the prospective new entrant, some process information is available in the public domain. Getting really detailed information costs money. You might hire a consultant to analyse processes for you, or you can try to do it yourself, remembering that many licensors will charge quite high 'look-see' fees for even partial details about their processes. In either case, you end up with information about a short-list of processes which has to be analysed in terms of the overall *project* resulting from the process selection.

Methods of project appraisal have been discussed in Chapter 9. They can be used when the project involves a technology licence, making due allowance for the risks associated with licensing as well as those associated with the technology.

11.4 Licence negotiations

LICENCE AGREEMENTS

A licence agreement is the outcome of negotiation, and will not, in any meaningful sense, be a 'standard' document. It will, however, embody ideas which are common to most if not all licence agreements:

1. The licence permits the licensee to use the licensor's *technology* to practise a *process* to make the *products* and sell them within a *territory*. The words in italic must be carefully defined.
2. The definition of 'technology' (and/or 'knowhow') normally includes reference to the licensor's patents, and consequently to provisions covering indemnity against claims of patent infringement.
3. Licensors want licensees to keep confidential the significant process information for a period usually of several years from the date of the agreement. This imposes quite a substantial burden on the licensee, especially if third parties such as contractors are involved.
4. The agreement stipulates the amount and terms of payment of the licence fee, either as a capital payment or as a royalty, usually based on output. In the latter case, a foolproof accounting procedure is needed.
5. Among the most important provisions are those which cover the supply of technology developed *after* conclusion of the licence agreement. Usually this is expressed in terms of an 'exchange' of developmental information, although often the exchange will be one-sided.

LICENSING AND CONTRACTORS

It very often happens that a licence will pass through a contractor, who is in fact the sole contract partner as far as the licensee is concerned. While the contractor will seek a back-to-back agreement with the licensor for the sake of self-protection, he may also hope to establish a preferred position in a particularly attractive technology, in order to sell more plants. The enthusiasm of licensors seldom matches that of contractors in this respect, and this highlights some of the problems of licensing.

The first question to be resolved concerns exclusivity. Usually process licensors are reluctant to grant exclusive privileges to any one contractor, although they may do so within a limited 'territory' if the contractor seems to have a dominant position there.

The licensor then undertakes to provide 'basic engineering' to the contractor. This information is usually defined as being 'of sufficient scope to enable the contractor to engineer and construct the contract plant'. The information is based on the licensor's own operating experience, and it is for the contractor to take account of his client's (i.e. the owner's) special requirements. Moreover, the amount and quality of technical information provided in the process manual can vary and does not necessarily correspond with what the client organization may need. Such discrepancies are a potential source of great difficulty for the contractors, who – in the nature of contractors – will aim to pass the problem on to their client.

Continuity

Licensing is often seen as a rather easy way to increase profits, although there is always the spectre of increasing competition as new licensees expand their market share, and what is won in the shape of royalties fails to match up to the licensor's expectation of profits from increased penetration of the market.

The stages by which a docile licensee evolves into an aggressive competitor are often complex and subtle. Examples of the souring of licensor–licensee relationships, despite initial goodwill, are most frequent when the licensed technology is continuously developing while the relationship is structured so as to be immutable.

Ideally, a licensor would seek to license a company which would be a 'good' competitor if – or when – it entered the competitive market. That is to say, the advent of the new player would strengthen the total market to the licensor's advantage, as well as that of the licensee. Again, licence agreements would ideally be subject to renewal clauses which limit the possibility of a perpetual commitment by the licensor (which, under the impact of unforeseen developments, might work to his disadvantage).

Perhaps the problems are psychological rather than technical or commercial, but the spirit of ongoing technology development and exchange withers if it does not find willing support from both sides (Brandt, 1980, 1983).

11.5 Joint ventures

TECHNOLOGY TRANSFER

Licences are common vehicles for technology transfer. A popular means of ensuring community of interest in the success of the transferred technology is the joint venture (JV). For our present purposes we can define a JV as a business whose ownership is shared between two (or more) partners, who normally also have other business interests.

Clearly it is unlikely that the partners' overall business interests will exactly coincide. But as far as the JV is concerned, they should converge and should continue to do so throughout the life of the JV in order to avoid conflict which can distort or disrupt the aims of the JV.

We may visualize the transfer of technology evolving in three stages:

1. transfer of products, i.e. development of the 'technology' of marketing and utilization in the transferee organization;
2. transfer of production facilities, i.e. development of the technology of production, operation and maintenance in the transferee organization;
3. transfer of innovating capability, i.e. development of the technology of locally applicable R&D capacity in the transferee organization.

Building a plant to use transferred technology (stage 2 above) is not, in itself, a transfer of production knowhow. It is an opportunity, limited in time, to achieve a certain competitive position which can be sustained only if the transferred technology contains and can be contained in other viable systems, which are themselves not in decline relative to the competition. Further, unless the transferee moves to the third stage, he discovers that his plant becomes obsolete – new technology has to be acquired. At that point, he has no option but to abandon the field or buy another ticket to remain in competition (Beaumont *et al.*, 1981).

It is this, the development of a lively, dynamic technological capability, which is – or should be – the motive for joint ventures, rather than (for instance) the opportunity to make money out of crafty manipulation of transfer prices. And it is agreement about this motive which underlies the licensing framework which we have been discussing.

There remain problems of day-to-day management which persist despite the most enlightened ideas. In many joint ventures, the licensor's participation is represented by capitalization of his licence fee, so that he has only a small proportion of equity. However, he inevitably remains primarily responsible for the success of the (technology-based) venture.

Because the success of the venture will be determined by its performance in the market, a licensor may seek:

- to control product marketing, either by contracting to take the major part of the output himself, or through a third party judged likely to favour his interests;

- to pass control of production to the licensee in return for control of marketing through a separate company controlled by the licensor;
- to have an effective majority on the JV Board, irrespective of his equity holding;
- to have the right to appoint executive management.

Not surprisingly, many licensees find these views unattractive, so that, in addition to the characteristics of licensors which we discussed at the beginning of this chapter, licensees should ask the following questions:

- Does the licensor inspire confidence that his present and foreseeable objectives for the JV converge with the licensee's own?
- Is the licensor willing openly to discuss areas of potential future conflict, especially on the distribution of profits and on expansion or other activities likely to call for the future injection of additional capital?
- Is the licensor willing to support and contribute to the licensee's efforts to develop the technology further?
- Is it possible to agree in advance principles for the equitable winding-up of the JV if that eventually proves desirable, even though considered unlikely now?

These often prove to be very difficult questions. It is, however, far better to address them early on in negotiations about the licence than to hope that they will, if left alone, go away.

JOINT VENTURE PITFALLS

Business objectives inevitably change with time, so that – at least in a dynamic business environment – the aims of the JV partners are likely to change. Then their business objectives will probably diverge. This tendency can be made worse (or, on the other hand, its effects reduced) in ways which depend on the partners' reasons for entering into a JV in the first place. Consider the following:

- 'We want to use the JV to share risks.' This is not a very good reason. It is difficult to divide up risks fairly, and you need to be single-mindedly in control to manage risk effectively.
- 'We want to avoid or minimize competition.' This means that the partners are likely to be at odds from the outset. There may also be statutory objections. However, sometimes an opportunity arises to buy part of an existing business, thus avoiding the risk that a competitor will buy it. The purchase may also be seen as providing access to useful 'inside' information. But it is important to be sure of the synergy of the proposed acquisition with our own business, and to recognize that expectations in this area are seldom matched by events.
- 'We need a partner so that we can exploit the technology in a place where we are prohibited from 100 per cent ownership.' This may be acceptable, provided we recognize we will probably lose control of the technology.
- 'We need a partner to put more money into the project so that we can exploit the technology to the full.' This is also possibly an acceptable solution but, if the project is successful, our partner may well aim to be independent, or if the project runs into difficulties, the JV will demand yet more money and our partner may grow fractious.
- 'We need a partner who has the market for the project's output.' This means we have to work out and agree in advance a pricing formula. No matter how carefully this is done, the chances are it will lead to conflict later.
- 'We need a partner who has knowhow (e.g. technical, production, distribution, marketing) which matches and complements our own.' This can be attractive, provided that future developments keep pace with business development.

Taking steps to moderate these difficulties may raise other potential areas of conflict. Money and control are frequent battle grounds.

Whose money?

Fiscal issues Taxation may affect the partners in different ways, so that they have different tax liabilities in respect of their stake in the JV. If one partner's tax liability turns out to be substantially different from the other's, conflict will probably ensue. Conventions in accounting may also differ, notably regarding the treatment of depreciation. Policy on reconciling such differences as these should be agreed early.

Financial issues Policy on the structure of capitalization and gearing must be agreed before launching the JV in order to avoid potentially serious damage later. Similarly, policy on dealing with the need for additional capital (e.g. for expansion or to meet unforeseen demand for operating capital) must be agreed. We should remember that cash in a JV is in effect locked into the JV company, so that the ability of the partners to move cash quickly from their overall corporate resources is limited.

Policy on payment of dividends is another important area of potential conflict which should be resolved early, and not – as sometimes happens – left to be argued over by lawyers at some later stage.

Who controls?

This can be tricky when each partner holds 50 per cent of the JV. Some 'licensor' approaches were mentioned in the previous section. Some other angles are:

- 'We issue shares of two kinds: voting and non-voting.' This distributes the JV's profits equally, but puts control in the hands of the voting shareholders.
- 'The majority of the shares is divided 50:50 between the JV partners. A small percentage is allocated to a third party friendly to us.'
- 'Shares are held 50:50. We appoint the management contractor.'
- 'If we are obliged to have 50 per cent local ownership in the JV, then we arrange for the locally owned shares to be held by a local company which has no interest in the management of the JV.'

It is said that human ingenuity knows no bounds. This is especially the case when one party seeks to dominate the other. If domination is the prime objective of a prospective partner, we must question whether his participation in the JV is really what is wanted.

Despite the difficulties, licensing thrives and does contribute to the profits of licensors and the benefits of licensees. It will continue to do so just as long as it is seen to be a reciprocal, and mutually advantageous, way of doing business.

11.6 Summary

A license is a way to acquire technology. The acquirer becomes a *licensee* (or technology transferee). It is also a way to exploit technology. The exploiter becomes a *licensor* (or technology transferor). If the licence is to be a vehicle for successful business, these parties have to establish some mutually acceptable objectives and motivations.

From the licensee's point of view, a prime concern is to acquire technology which is not now, and is not likely to become quickly, obsolete. The choice of technology therefore depends on some form of forecasting and on negotiation with potential licensors to establish a long-term relationship which should provide for ongoing technological support.

Technology licenses are often acquired through a contractor (e.g. associated with a turnkey project contract). The advantage that the owner then has only one party to negotiate with may, however, be negated by the difficulty of finding out who is responsible if the project fails to perform to specification.

A joint venture is a popular way of establishing mutual interest between the transferee and the transferor of technology. Mutuality of interest is, however, vulnerable to attempts (by either party) to bias control of the JV. The agreement which sets up and structures the JV must be carefully drawn so as to address these problems from the outset, rather than ignoring them until they have become crises.

12
Consultants and Contractors

12.1 Consultants

CONSULTANCY AS ADVICE

> The service provided by an independent and qualified person or persons in identifying and investigating problems concerned with policy, organization, procedures and methods; recommending appropriate actions and helping to implement these recommendations.
>
> (Institute of Management Consultants)

> A chap who borrows your watch in order to tell you the time.
>
> (Client's folklore)

There is enough similarity between these descriptions of what consultancy is for us to be able to see that it covers a wide range of interesting activities. The origins of consultancy go back to soothsaying,* which simply meant having special expertise or insight enabling one to give a truthful (or at least convincing) account of the future. In a word: *advice*.

Advice comes in various categories. Modern consultants may:

- provide information which is not easily available;
- interpret the information, so as to provide an impartial, objective point of view based on their experience;
- provide special knowledge and/or knowhow;
- provide arguments which justify measures that managements have already decided upon;
- supply temporary help.

In all cases, the advisory services which consultants provide are generally to do with 'problem solving'. If the problem is known – or rather, if the structure of the problem is known – problem solving becomes a matter of applying appropriate techniques with relevant judgement. This means that 'competence' – the 'combination of knowledge, skill and attitude that enables optimum performance' (SIPM 1993) – together with 'experience', are necessary attributes of consultants.

* 'Sooth' originally meant 'truth', but has linguistic connections with soothing pain and with flattery. In former times, soothsaying covered everything from being high priest to being court jester. Modern consultants may be thought to offer a similar range of services.

In the best consultants, they add up to *insight*.

'Insight' is a difficult concept to explain. It is discernment, penetration, understanding, and it is perhaps best shown by the problem illustrated in Fig. 12.1: what is the shape of a single plug to fit exactly each of the three holes (Perelman, 1973)?

Fig. 12.1 Insight?

CONSULTANCY AS INFORMATION

Obviously, a main plank in the provision of advice is information. Consultants provide information in a number of ways:

- **Published information.** This means information which is in the public domain, but which may not be readily accessible to clients in the time available. Most consultants maintain their own databases, and can link into national and international databases quickly. Perhaps more important than these in areas of specialized information, consultants usually cultivate personal contacts which they hope can be mobilized to provide the key pieces in the information jigsaw puzzle.
- **Multiclient studies.** Many consultants produce multiclient studies which provide data on specialized topics at a lower cost to individual customers than would be possible for a specially commissioned study. Multiclient studies can help client organizations to build their own databases, but they do need to be treated with caution. While they provide background information, they evidently contain nothing about any particular client's situation, and consequently may be misleading, especially in the context of market competition. Sometimes, the updating of multiclient studies is patchy, and this adds to the risk of using them in an uncritical way.
- **Specially commissioned studies.** This is by far the best way to get 'insight-full' information on any significant problem. The only disadvantage is that specially commissioned studies are expensive. Because it is the small but crucial details which take most time and effort to obtain, clients frequently try to save money on specially commissioned studies when a multiclient study in what is nominally the same field is available. However, the chances are that they will have to spend even more rectifying the mistaken decisions they have taken on the basis of inadequate information.

Practically all information for use in the client organization has to be interpreted, in the sense that the consultant has to analyse, synthesize and summarize information in the context of the client organization's interests and aims. It is important that the activities of analysis (e.g. breaking down the data available to discover what is salient, what conditional), synthesis (e.g. construction of arguments, propositions and alternatives) and summarizing (which includes comparison and appraisal) should not become confused: 'that way madness lies'!

It is at the interpretation stage that problems with information become apparent. The main problems are:

- insufficiency – the information is incomplete, and so provides less of a basis for development than is necessary;
- unsuitability – the information is either irrelevant or simply wrong, so although it appears to be complete, it is flawed;
- bias – this is a special case of wrong information, in that it is misleading due to accidental or deliberate causes.

Simply reviewing these information problems shows the importance of independence in consultancy. Information – and the advice based on it – is quite useless if it is merely what the client wants to hear, or if it disguises a vested interest.

CONSULTANCY ASSIGNMENTS

Clients use consultants in many ways. In addition to getting and interpreting information, consultants may also be retained to make recommendations for clients' decisions and actions and to help implement them. A typical example is the preparation of plans and the formulation of a strategy for development of the client's business. Clearly, this may include project development.

Within this framework, examples of typical assignments include:

- **marketing research** – where, how, how much to sell, and to whom; acquisitions, divestments, diversifications;
- **commercial research** – pricing policy, transfer pricing; competitor analysis; contract and licence terms analysis;
- **technical development** – technology evaluation; organizational planning; plant revamps and de-bottlenecking.

This list is far from exhaustive. A typical feasibility study is likely to contain elements from all the items mentioned above and may include others as diverse as quality management, environmental planning or energy conservation.

Considering consultancy more specifically related to project leadership and management, consultants may be assigned to:

- perform critical reviews of project feasibility studies;
- review the availability and reliability of supplies of crucial inputs to the project (e.g. utilities, special consumables etc.) and the logistics of disposing of outputs;
- review technological options in relation to, for example, potential obsolescence, concurrency (technological uncertainty + time pressure), manpower availability;
- review energy use options;
- plan project organization and contracting strategy in relation to project financing options;
- short-list contractors and prepare bid specifications and invitations to bid (ITBs);
- check and compare contractors' bids for inconsistencies, conformity with ITBs, scope changes and general quality;

- advise owners on commercial terms proposed by contractors, and assist in obtaining final price revisions and contract terms;
- supervise contractor's design, purchasing, construction and commissioning activities.

CHOOSING CONSULTANTS

Whether you are looking for an individual consultant or a team, you have a hard choice to make, and one where, if you get it wrong, you may cause major problems for your company. What can you do to make as sure as possible that your choice is the right one?

First of all, perhaps we should clear up the question of size. Clients, especially those who are unsure of themselves, often believe that they have to choose 'big name' consultancies, meaning those that have grown so large that their fee income gets quoted in newspaper league tables. From the client's point of view, this may offer a let-out if (or when) the consultant's advice proves to be wrong as well as expensive. There is in fact no inherent advantage to the client in the size of the consultancy. The client is paying for the quality of advice – and that relates only to the quality of the consultants working on the assignment, as individuals and as a team.

The Management Consultants Association has produced a set of guidelines, which is the basis of the following observations. They all concern the *client's* attitudes of mind: if the client understands his side of the assignment, the chances are that the assignment will go well and be productive; if he doesn't, it almost certainly won't:

- Accept that no consultancy can do everything. Any one that offers to do so should be treated with suspicion.
- Understand that consultancies are also businesses. Find out how the consultant proposes to manage his business while he is advising yours.
- Meet the key people in the consultancy who will be responsible for your assignment, and make sure the people you meet are in fact the people who will do the work.
- Be prepared to be quite open and frank in discussing the assignment with the consultants. Expect the consultants to be equally open and frank with you.
- Realize that the only assignments which are free from conflict are trivial ones. In particular, several people on your side – including you yourself – may resent what the consultants are doing. If you are not ready for this, you are not ready to place the assignment. If the consultant is not ready for it, he is not fit for the assignment.
- Avoid open-ended assignments, which are bad both for the consultancy (which loses objectivity) and your own people (who get to be dependent on the consultants). One way to deal with large, ongoing assignments is to break them up into separate phases. Each phase is separately costed, and the next phase is authorized only on the basis of the progress demonstrated in the current phase.
- Don't accept that the cheapest quotation is necessarily the best offer of consultancy services. Price alone is not the best criterion of choice for most things. It has to be taken into account of course, but in the case of the consultancy, it weighs rather little in comparison with factors such as capability, relevant experience and track record, each of which, however, is extremely difficult to measure. This simply means that you are going to have to do some hard work in selecting consultants.

Client pressures and the other forces of competition tend to lead both parties to underestimate the cost and time needed for consultancy assignments. Warren's Rule (a corollary of Murphy's Law) says: 'To spot the expert, pick the one who predicts the job will take the longest and cost the most.'

The best way to reach a reasonable balance between pressure of time and pressure of cost is whole-hearted commitment by the client to the aims of the assignment. Sometimes this takes the form of a client's project leader (perhaps with his own team) who runs the assignment on the client's side; the secondment of one or several client's people to the consultancy team; or simply providing the consultant's team leader with a company 'anchor' person who can set up meetings and generally get things done within the client organization.

Defining the assignment

Consultancy assignments tend to stretch, like elastic. It is important to define them in such a way that this property can be utilized when it is advantageous to do so, but is rigorously controlled otherwise. For most assignments, this is best done by preparing terms of reference for the assignment, which eventually become the basis of inviting detailed proposals from a short-list of consultants.

The task of preparing terms of reference is, in itself, quite difficult. Sometimes, clients assign this work to consultants (who then must forego bidding for the assignment to which the terms of reference apply). Whether undertaken in-house or assigned to an outside consultancy, a useful approach to writing terms of reference is to treat this like an engineering job: define the battery limits of the project (i.e. the scope of work in relation to its concepts and objectives in time and space) and its tie-in points (i.e. where it interfaces with the rest of the organization and the outside world).

The terms of reference should contain enough information for the bidders to be able to show:

- how they propose to undertake the assignment;
- who they propose to do it;
- the proposed programme of work and schedule for delivery;
- the quotation for fees and for other costs;
- what their requirements are from the client organization;
- any differences in their concept from the terms of reference, their reasons, and how they propose to accommodate these differences.

It is in fact quite rare for honest consultants to have absolutely no difficulties in meeting clients' original terms of reference, so it makes sense to set up meetings with the possible contenders for a consultancy assignment to discuss all and any issues raised by the terms. Needless to say, clients must be prepared to provide complete information at these discussions.

Short-listing

Short-listing implies an obligation on client organizations to know enough about prospective consultants to be able to construct a short-list for any likely assignment. Client organizations are notoriously lazy about maintaining this kind of information. This explains why a large proportion of assignments are wasted or, at best, show poor value for money. However, supposing that proposals are received from a properly drawn-up short-list of consultants, how should one go about evaluating the proposals?

Here are a few basic criteria:

1. Does the proposal show that the consultants understand the terms of reference? There are three levels on which this should be judged:
   (a) with respect to the assignment;

   (b) with respect to the client organization, and in particular its 'business culture', i.e. the opinions and attitudes of the client organization's decision makers which condition (though they may not determine) the way the client looks at its own business;
   (c) with respect to the 'external environment', i.e. the multiplicity of factors which influence the client's business without being in any way controllable by the client.
2. Does the proposed scope of the assignment correspond with the terms of reference? If not, why not? It may be that the consultants have very good – and acceptable – reasons for proposing work that does not precisely correspond with the terms of reference. These reasons have to be made clear, and their consequences explained, for otherwise it will prove difficult, if not impossible, to monitor and control the consultants' work.
3. Is the proposed methodology and the programme of work reporting consistent with the proposed scope of the assignment and with the client's terms of reference (modified if need be)? In other words, if the consultants cannot show how they will undertake the assignment and explain their approach, it is not very likely that they will be able to carry it out satisfactorily.
4. Is the proposed consultancy team credible, i.e. are the qualifications and experience of the proposed key members relevant and appropriate? Neither qualifications nor experience need be identical to the client's supposed requirements, but they should provide credible evidence of capability in the field under study. It is especially important to be sure who will actually do the work. Most consultants can produce the CVs of a few distinguished individuals: what the client needs to know is exactly who will be attending to *his* problem.

   In this general area of credibility, consider also the personality of the key team members, especially of the proposed team leader. He may be brilliantly qualified and broadly experienced, but if his personality makes it impossible for your team leader to work with him, you will have to change one or the other – or look elsewhere.
5. Is the consultancy firm credible? This question mainly concerns formal responsibility for satisfactory performance of work. It is more to do with management experience of consultancy assignments (which is categorically different from management of, say, product marketing or plant construction) and the organizational capability of the firm. For example, how will it cope with the unavoidable non-availability of one or more of its key team members? The consultancy firm's credibility also depends on its background experience of working with similar clients – even your own competitors. In addition to the consultant's own representations, their credibility should be ascertained from other references – don't rely on what may appear in consultant registration documents.

Commercial terms The consultant's proposal will (or should) contain references to items such as 'confidentiality', administration of the assignment and the way in which any disputes should be managed. In practice, most client organizations are so much more powerful than most consultancies that the clients can always dictate these terms if they wish to do so. But the consultants' proposals in these respects are valuable pointers as to their attitude to the job. If the consultants demonstrate practical common sense in these areas, there is a good chance that they will handle the assignment in a similar vein.

Fees and costs We have already discussed the pros and cons of low quotations. Basically, there are very few ways in which fees and costs can be proposed, and the essential guidelines are as follows:

- If the scope of the assignment can be reasonably clearly agreed, the consultancy should be prepared to quote a firm fee.

- If the scope of the assignment cannot be defined, the client should be prepared to pay fees on a per diem rate basis. But this should be converted to a firm fee basis as soon as agreement can be reached on definition of the scope of the assignment.
- In very few cases (typically long-term strategy development assignments or assignments related to training), it makes sense to agree a retainer with consultants. The objective of a retainer is to give clients a priority claim on the availability of consultants when both the scope and the timing of the assignment are in doubt. When these unknowns are defined, the assignment should be brought within the framework of a firm fee or per diem rate agreement. (In other words, a good consultant should be trying to work himself out of a retainer agreement, despite its obvious attraction!)
- Expenses (typically for travel, accommodation and subsistence, but sometimes including items such as communications, computer time, and purchased services) should be reimbursible at net cost to the consultant. Clients who insist that these expenses should be borne by the consultant within his total price will pay more than they need. These expenses are not within the consultant's control, so if he has to provide for them, he will add in a contingency to cover the risk.

Clients should appreciate that consultants are not in business to provide financial credit. The phasing of payments to consultants therefore has to ensure that the assignment is effectively self-financing. If clients insist on a schedule of payments which undermines this principle, they, the clients, will eventually pay the cost, which is likely to be considerably higher than the cost of money to finance proper payments (Fig. 12.2).

Fig. 12.2 Payment by the job.

Selection from the short-list

In the previous section, we set out some criteria for short-listing the bidders for a consultancy assignment. Suppose we are now going to assess the bids and choose the best. How should the various criteria be weighted?

This will depend very much on how the client organization sees its own part in the assignment, for consultants rarely work in total isolation from the client. As an illustration, the following weights are fairly typical in the case of an 'arm's length' assignment for an enlightened client:

- Does the bid show that the consultants properly understand the terms of reference? (Maximum score: 15 per cent.)
- Does the proposed scope of work match the terms of reference? If not, are the reasons properly (and acceptably) explained. (Maximum score: 15 per cent.)
- Are the proposed methodology, schedule of work and arrangements for reporting sensible? (Maximum score: 15 per cent.)
- Are the members of the proposed consultant team suitable, e.g. qualifications, experience, personality? (Maximum score:15 per cent.)
- Is the consultant firm credible, e.g. credentials, organization and management, back-up resources, relevant background and ownership? (Maximum score: 10 per cent.)
- Contract proposals, e.g. confidentiality, administrative provisions, resolution of disputes, etc. These are the parts of the bid which reflect the 'enforceable' aspects of the consultants' attitude to the assignment. (Maximum score: 15 per cent.)
- Price – fees, expenses and the phasing of payments. Are these appropriate to the nature of the assignment and the rest of the consultants' proposals? (Maximum score: 10 per cent.)

This procedure is by no means universally accepted. Some clients adopt a two-stage evaluation process in which (1) consultants' technical submissions are evaluated and ranked in order of technical acceptability, and then (2) a separate evaluation team negotiates the price with the best technical bidder, and – if that negotiation fails – tries again with the next best technical bidder and so on until a satisfactory agreement is reached.

Another version of this procedure is to weight the technical part of the consultant's bid (e.g. understanding of, and compliance with, the terms of reference, methodology, organization, team, firm's capabilities and experience) separately from the financial part (e.g. price). The technical part is rated as a percentage score against criteria such as those mentioned above. The financial part is rated in relation to other bids:

Financial rating of firm A = 100 × lowest price/firm A's price

Typically the technical rating is weighted 70–80 per cent; the financial rating is weighted 30–20 per cent. The 'best' offer is taken to be the highest weighted average of the technical and financial ratings.

MONITORING CONSULTANTS' PERFORMANCE

It is common sense for any buyer to take care that what he buys corresponds to what he was offered when the deal was struck. Clients for consultancy services often fail to do this, or alternatively try to monitor consultants' performance so closely that they actually interfere with the work in hand.

In any substantial assignment there should be provision for the consultants to report on the status of their work at reasonable intervals. In short-term assignments, interim reports may not be feasible, but even then it is sensible for clients to see a draft report and discuss it with the consultants before the final report is completed. This is not an occasion for browbeating consultants into changing their minds, or for leaping to premature conclusions. It is an opportunity to review data, and particularly to bring out into the open any assumptions which need to be made explicit.

Interim reports are a nuisance and should be kept to a minimum. They prejudice the formation of well-founded conclusions and they waste the time of consultants who should be doing serious work on the assignment. A better way for clients to keep abreast of progress is to hold fairly frequent progress meetings at which the consultant's project leader presents a verbal status report (with perhaps a simple one page written summary) stating what has been achieved to date and what the aims are for the period up to the next progress meeting. Phased payments are often linked to performance as marked by these progress meetings. This helps to ensure that they are taken seriously by the consultants.

No respectable consultant wants to be associated with failure or poor value for money. No respectable consultant will therefore object if you want to monitor his work – although he will certainly object, and tell you so loudly, if your attempts to monitor get in the way of proper performance. The sensible approach is cooperation, not confrontation; this is perhaps the most important of all criteria affecting your use of consultants.

12.2 Contractors

CHOOSING CONTRACTORS

Contractors, like consultants, live by selling manhours. But whereas consultants' manhours represent what is essentially information and expert advice, contractors' manhours cover a wide range of services connected with 'doing' rather than 'advising' about the project. Contractors typically 'do' the following: design, purchasing (of equipment and materials, and sometimes other contractor's services), construction (of the facility) and commissioning (starting up operations and demonstrating performance), and they may be involved in, for example, arranging finance, training the owner's staff, and even running the facility as a business operation.

Just as the projects on which contractors are engaged range from multibillion dollar international schemes to small structures costing a few thousand dollars, so contractors themselves range from huge internationally active firms to small local enterprises. The technical departments of manufacturing companies often function as contractors in their parent organizations, and they sometimes evolve into profit centres capable of exporting their manhours to the world at large.

The processes of choosing and using a contractor are similar to those for a consultant. When, in the previous section, we discussed the latter, our aim was to choose the party which could best *advise* us about our project. When we choose a contractor, our aim is to choose the party which can best *manage risk* in our project. Apart from this, the procedures for short-listing prospective bidders, evaluating their bids and making a final choice will be quite similar. In brief, we will:

- prepare a bid list of suitably qualified contractors;
- send to these contractors (and to no others) an invitation to bid (ITB);
- make a 'fair price' estimate as a datum for bid comparison;
- evaluate the bids on a strictly comparable basis.

Choosing a contractor is perhaps a little easier than choosing a consultant, because the information available for evaluation will be a little more tangible. The consultant's fee is likely to be much less than the contractor's contract value, but, on the other hand, the consultant's advice, if wrong, will probably hurt you more than a failed contract.

THE 'INVITATION TO BID' (ITB)

A typical ITB will contain:

1. **Introduction** – enough general background information about the business environment of the proposed project for the contractor to understand its purpose. Reference should be made

to all relevant statutory instruments, standards, codes of practice etc. which affect the project works.
2. **Bidder's credentials** – contractor to provide details of recent track record, current and projected workload, and key project personnel.
3. **Objectives of the project** – what it is intended to achieve, so that the contractor can develop appropriate design and implementation philosophies.
4. **Design basis** – the contractor must conform to this (although he is permitted to offer alternatives) so it should be clear and unambiguous.
5. **Scope of work** – specifies what the contractor is required to undertake, in terms of project deliverables, in order to realize the project objectives. Stipulations regarding the use of local partners, subcontractors etc. should be specified here if applicable.
6. **Implementation plan** – invites the contractor to show how the project work will be organized and carried out. Should include the quality management plan, and procedures for reporting project progress to the owner.
7. **Project schedule** – contractor's detailed proposals for the timing of project activities, including the owner's involvement.
8. **Financial proposal** – contractor to quote prices/rates on a specified basis (e.g. lump sum, reimbursible, etc.) and proposals for credit or other financing if applicable.
9. **Terms and conditions** – either specified in detail (or by reference to a specific standard form of contract) by the owner or requested to be proposed by the bidder. They should include proposals for incentives/penalties and the method for delivery of the bid (e.g. separate sealed technical and commercial proposals), delivery addresses and deadlines. They should indicate how the bids will be assessed, and when the results of the assessment will be notified.

A bid designed to meet the requirements of such an ITB will probably be set out in the following sections:

- the contract proposal, including prices and terms of payment;
- technology licence proposal, if applicable;
- technical annexes covering:
  - design basis
  - details of specific performance
  - process description
  - hardware (equipment and materials) to be supplied by the bidder, including construction equipment and spare parts/consumables if applicable
  - responsibilities of the owner
  - documentation to be supplied by the bidder
  - documentation necessary to be supplied by the owner
  - schedule of deliveries and project work
  - performance guarantees undertaken by the bidder
  - services to be provided by the bidder.

If the ITB contains sufficient clear information, the contractor will feel confident about his ability to make a realistic bid. Otherwise, depending on how hungry he is for the job, he will seek clarification from the client or choose deliberately to submit an unacceptable bid (disguised as a gesture of goodwill). Alternatively, of course, he may abandon further effort.

If the contractor does decide to make a bid, he will aim to ensure that:

- the response is appropriate to the client's requirements;
- it does not leave out important cost elements, either through lack of information or from carelessness;
- allowances and contingencies are kept to a minimum;
- the expenditure of resources in making the bid is sensibly related to the profit expectation if the bid is successful.

BID EVALUATION

The most important instrument for bid evaluation is a 'fair price estimate' (FPE). This is an estimate of the cost of carrying out the project which the client makes (alone, or with the help of consultants) in essentially the same way as the contractor prepares his bid.

However, using the FPE comes at a later stage. We need first to consider several other factors. It helps to set up a checklist of contractor assessment criteria. A typical list would include:

- ownership, including parent/subsidiary relationships;
- management structure, with main lines of responsibility and communication;
- staff numbers and disciplines;
- proportion and assignment of temporary vs permanent staff;
- personal résumés of key people;
- work capacity of the main departments;
- current and projected work loads;
- recent contracts completed – experience and track record;
- location and type of offices and other facilities;
- financial resources and status.

This list is not exhaustive. It does, however, show that collecting the information on which to base even a preliminary assessment will take time and a lot of effort. It is therefore worthwhile to set up a data bank on contractors for future reference, updating the information periodically and reviewing the data bank as a whole, say, every 2 years.

By weighting the checklist criteria to reflect their relative importance and awarding scores to each contractor based on the contractor's bid submission, we can rank the contractors according to their weighted scores. Since both weights and scores are inevitably subjective, it is usual to get three or four evaluators to make the assessment independently. The result will usually be substantial agreement on the contractors which should *not* be short-listed, and debate on borderline cases. It may be necessary for the client's project team to visit the contractors so as to satisfy themselves regarding the information provided in the bids. The result of all this is to generate a suitably short-list of potentially acceptable bids for detailed negotiation.

Consortia of contractors need extra care in assessment. In addition to the previous list of assessment criteria, we need to assess:

- whether the members of the consortium are well matched;
- each member's responsibility in the project;
- the position if one (or more) member(s) fail to meet their obligations;
- whether the members are 'jointly and severally' liable for the performance of the contract.

MAKING THE FINAL CHOICE

The final choice is normally made from the short-listed contractors whose bids have been received by the due date and according to the conditions specified in the ITB. The 'negotiated bid' (where only one bidder is considered) is a special case, usually where a particular contractor has very specialized skills or is already working satisfactorily on the site. Occasionally, it may be decided that a project is too small and too specialized to justify the expenditure of time and cost involved in competitive bidding, and in this case a negotiated bid is preferred.

Competitive bidding demands that the competitors bid on the same basis. It is otherwise not possible to make a fair comparison. However, contractors are always free to propose alternatives. They should, however, be required to comply with the bid documentation (including alternatives) and, if an alternative is put forward, to explain the reasons.

Selection is based on procedures similar to those described for consultants (p. 160–1), taking account of commercial and financial proposals and political factors as well as the technical and organizational factors. Very careful consideration should be given to the contractor's compatibility with the owner at project team/project manager/project director level. If these individuals cannot establish good working relations, even the best contract strategy is likely to fail. If they can, many otherwise serious problems will disappear.

Compatibility

Compatibility has much to do with ensuring good communications. Many experienced project managers consider good communications to be the most important single factor in running a successful project. It is easy to see how difficult it may be to achieve good communications in a project if you consider the flow of information between those who originate it and those who use it. The number of interfaces can grow very rapidly, and so does the number of opportunities for information to go wrong (Chapter 7)!

Price

The bid price is not, in itself, a measure of the contractor's capability for managing project risk. It is inseparable from the other terms of the bid. The basis for these will have been set out in the owner's ITB and, as we have seen, part of the evaluation process is to determine how far the bid complies with the ITB (or, if it does not, whether the reasons for non-compliance are justifiable; note that if they are justifiable, the owner may wish to revise the ITB so as to ensure that all the bidders re-compete on an equal footing).

Assuming that the bids do comply with the ITB, i.e. they cover the same scope of work and extent of supply, the bid prices may differ for a variety of reasons. Among the most common are:

- something has been omitted from the bid price;
- allowances and/or contingencies have been wrongly estimated;
- financing proposals are inappropriate.

The reasons for price differences may be identified by the owner only by using his fair price estimate (FPE) as the yardstick for checking the contractors' bid prices. If something has been omitted, it may be due to a misunderstanding, carelessness or deliberate intent. In any case, it is better to clear up the reason in discussion, and not to try to take advantage later. (Deliberate omission of items in the bid price may be construed as an attempt to profit later from 'variations to contract' or change orders, and should be dealt with accordingly!)

Allowances and contingencies are the contractor's confessions of ignorance, sometimes about circumstances where he has insufficient information (and where the owner should be willing to provide, if he can, additional information so as to reduce the allowance), and sometimes about circumstances where his experience leads him to expect some additional costs (quite usual – only the amount of the contingency may be disputed). Often, however, the total of allowances and contingencies includes double (even triple) counting, as occurs when engineering managers add their percentages to the estimator's, the project manager adds his percentage, and the managing director then adds *his* on top of the others. Comparison with the FPE helps to pinpoint these signs of risk management *non*-capability.

Contractors' bids often include proposals for financing the project. From the owner's point of view, the total price (i.e. including the cost of finance) is one thing, while the flow of funds to the project is another. Funds are required to match the phasing of project expenditures. If they do not, the project schedule is at risk. If they are more generous than the project requires, the contractor is financing something other than the project, almost certainly at an eventual cost which is higher than necessary. In some cases – e.g. export credit finance – the owner may gain from a disguised subsidy, but this has more to do with assessing the motives of politicians than with the merits of the contractor.

Payment terms

The baseline relationship between cost and schedule is shown by the 'S' curve throughout the project implementation phase. When payments to contractors are related to work performed the shape of the 'S' curve will reflect the payment terms. For example, the project schedule may be based on 'early' or 'late' dates for project activities in the project network. Payments may be made in proportion to work done (i.e. on a 'level of effort' basis) or, for example, 50 per cent on commencement, with 50 per cent on completion, of milestone activities. The choice will influence the baseline curve, possibly quite considerably (*see* Fig. 12.3).

Contractors are always eager to be paid as early as possible and to carry out their work as late as possible. Their initial proposals will be based on this approach. On the other hand, owner organizations will aim to have project work completed as early as possible, but to pay for it as late as possible. The difference between these approaches often leads to adversarial negotiation, i.e. to confrontation.

There are three fundamental pitfalls on the road to developing harmonious contractor–owner relationships; first, doubt concerning the contractor's ability to provide the effort actually called for by the project contract; second, the owner's difficulty in choosing a contractor with the skills

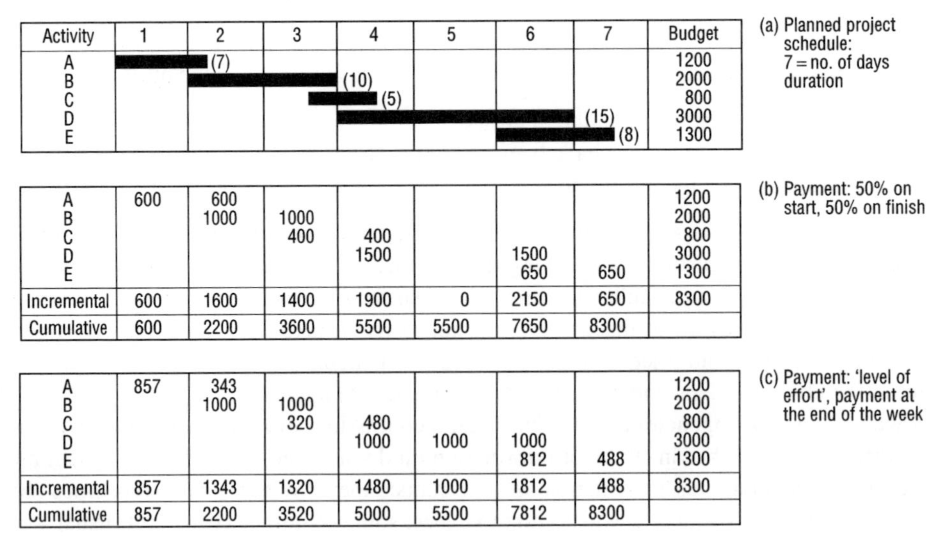

(a) Planned project schedule: 7 = no. of days duration

Activity	1	2	3	4	5	6	7	Budget
A		(7)						1200
B				(10)				2000
C				(5)				800
D							(15)	3000
E							(8)	1300

(b) Payment: 50% on start, 50% on finish

	1	2	3	4	5	6	7	Budget
A	600	600						1200
B		1000	1000					2000
C			400	400				800
D				1500		1500		3000
E						650	650	1300
Incremental	600	1600	1400	1900	0	2150	650	8300
Cumulative	600	2200	3600	5500	5500	7650	8300	

(c) Payment: 'level of effort', payment at the end of the week

	1	2	3	4	5	6	7	Budget
A	857	343						1200
B		1000	1000					2000
C			320	480				800
D				1000	1000	1000		3000
E						812	488	1300
Incremental	857	1343	1320	1480	1000	1812	488	8300
Cumulative	857	2200	3520	5000	5500	7812	8300	

Fig. 12.3 (a), (b) and (c).

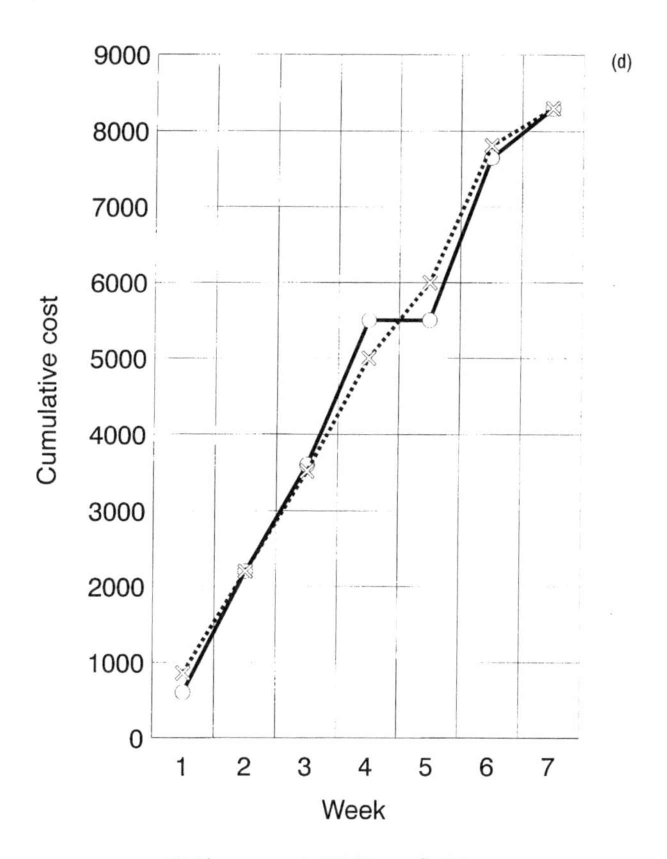

Fig. 12.3 Payment terms. We have a project of five activities, overall duration 7 weeks, with the planned duration of each activity shown in days in Fig. 12.3(a). In Fig. 12.3(b) we plan to pay 50 per cent of the activity's budgeted cost at the start, and 50 per cent at the finish of the activity. In Fig. 12.3(c) we plan to pay in proportion to the work on an activity completed during the respective week (accounting period). The 'S' curves for the different terms of payment are plotted in Fig. 12.3(d). These are planned (or 'baseline') curves. If progress of work does not correspond with the plan, different S curves will result. Variations from the baseline are the basis for control (see 'Earned value analysis', p. 211).

actually demanded by the contract; third, the motivations of both parties in the project situation where the contract is, in effect, the instrument for sharing risk in the project.

The getting and use of information intended to control the contractor account for a great deal of the effort expended trying to avoid these pitfalls. But the scale of payments and the method of making them are the main means of motivating contractors. Moreover, the payment terms can be an important aspect of contractor selection.

In deciding the terms of payment, there are two fundamental cases: a *fixed price* contract and a *reimbursible* type of contract. Both have advantages and disadvantages. On balance, owner organizations prefer a fixed price arrangement, where the scope of the project work is fully defined and agreed. The reimbursible approach is preferred in situations where the project work scope is not – or not yet – fully specified, although other considerations must also be taken into account (see Chapter 13).

However, *incentive* contracts, which may be based on either fixed price or reimbursible concepts, offer interesting possibilities of risk sharing on the basis of a rationale which should be more appropriate than mere bargaining. The rationale depends on:

- identifying categories of risk which are controllable by the contractor;
- identifying categories of risk which are controllable by the owner organization;
- identifying those categories of risk which cannot be controlled by either party.

Different payment arrangements should be agreed for each of these categories of risk. But this seems unlikely to be feasible unless the parties agree that cooperation is of greater mutual benefit than confrontation (Ward and Chapman, 1994).

As with consultants, contractor's work must be monitored: 'You get what you *inspect*, not what you *expect*!' The scale of inspection required in a typical contractor project is of course much larger than that for a consultancy assignment, and it is sometimes difficult for an owner organization which is relatively new in the field to put together a team of sufficient experience to monitor a major contractor's work. In this case, the owner will usually do far better to retain consultants to act on his behalf than to give the contractor a completely free hand.

12.3 Summary

Consultants provide services which are essentially in the nature of *advice*. They supply information and may manipulate the information to make recommendations about, give guidance on, or supervise their client's business. Despite the claims sometimes made on their behalf, consultants cannot do everything, are not omniscient, and do need clear terms of reference for their assignments. In specifying a consultancy assignment, clients must give the most complete background data.

Consultants should be selected on the basis of their compliance with the terms of reference for the assignment. Clients should establish suitable criteria and use a transparent selection process to choose from short-listed consultants in as open and objective a manner as possible. This also assists in monitoring consultants' performance in their assignments. Reporting is a necessary part of monitoring consultancy assignments, but it must be organized carefully so as to be effective in demonstrating real progress.

Contractors provide services which are essentially to do with putting the components of a project together. The range of contractor organizations matches the range of project types, from very large, hi-tech, international operations to small, basic, local businesses. The process of choosing a contractor is quite similar to that of choosing consultants. Usually the procedure is rather more formal because of the amount of money at stake, but it is perhaps a little easier because information about the candidates is usually more firm. Contractors bid against an invitation to bid (ITB) document, analogous to the consultants' terms of reference. Selection is based on ranking bids for compliance with the ITB, using suitable criteria. The owner's fair price estimate is an important instrument for checking bid prices.

Payment terms and contract strategy are usually more important in choosing contractors than they are in consultancy projects. However, in both cases, the aim should be to achieve a framework for cooperation, rather than confrontation, ensuring at the same time that performance can be properly monitored.

13
Contract Strategy

13.1 Introduction

A crucial stage in the project cycle is 'negotiation'. It is the stage at which all previous preparatory work is brought together to realize the project. The means for achieving this include many kinds of instrument intended to govern the way in which the parties involved in the project will work together. These instruments go under a variety of names, such as agreement, license, contract, and so on. Here, I shall refer to them all as 'contracts', because that is their *functional* purpose. We shall examine them from the functional point of view of the leader or manager of projects. There may, however, be different *legal* implications, depending on the particular circumstances of a project, about which you will need guidance from your company's legal advisers.

The terms of contracts are established by negotiation. This is a process by which we may suppose (perhaps idealistically) that the parties seek to decide, by reciprocal concession and compromise, procedures for optimizing their working relationship (*see* Fig. 13.1). In as much as the parties intend the instrument to be legal and enforceable, it will have some formal characteristics, but because it will apportion obligations and liabilities, risks and rewards, it will also have the characteristics of a mechanism for motivating the parties.

It is often said that the function of a contract is to assign project risks to the party best able to manage them. This is a platitude sufficiently bland to soothe the self-esteem of any cowboy jerry-builder, yet it contains enough truth to give the flavour of good sense. We need to think how to

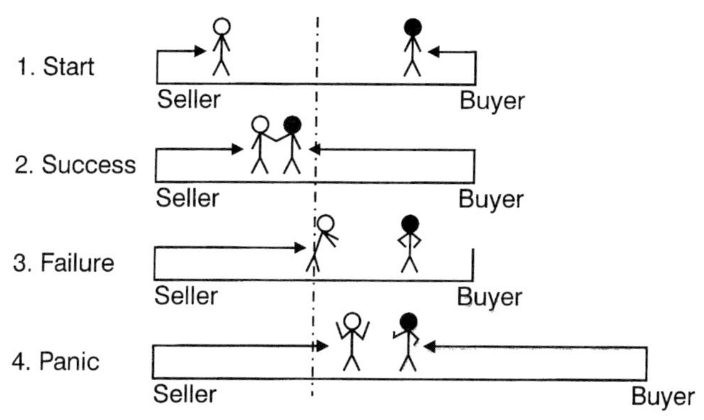

Fig. 13.1 Reciprocity, concession and compromise.

employ this good sense in the management of our projects, in much the same way as, during earlier phases of project development work, we tried to define strategies (i.e. consistent and coherent systems of ideas) to underpin our planning. We need to evolve a contract strategy too, for this will guide the negotiations by which we establish the 'legal and enforceable' framework for cooperation in work on the project.

In a large project, the amounts of money subject to the provisions of the contract are likely to be huge. But the contractual arrangements between the parties also display their respective corporate images to the public. For example, a major contract which encourages the employment of local suppliers in a project for a developing country will often establish a favourable image for the main contractor.

The influence of the contract strategy is far-reaching. For this reason, contract strategies should be considered early in project development, and care should be taken that the strategies are agreed with corporate management so as to avoid the possibility of conflict with other corporate interests.

In this chapter, we will set out the main features of contract strategic thinking, not as lawyers, but as managers recognizing that contracts are the means through which the project's 'fitness for purpose and economic use of resources' are actually brought about.

Throughout this chapter, everything is stated in the context of English law. Case precedents are very important in English law, and examples can often be found where precedents contradict the obvious interpretation of the case. (For example, the date of coming into effect of a contract has been held to be the date of posting the letter of acceptance, even though that letter never arrived. This kind of situation provides bread – and a good deal of jam too – for lawyers.) Evidently, other codes may lead to other interpretations, which may be very significant factors in a contract strategy, though the basic concepts in setting up a strategy seem to be international in character.

13.2 What is a contract?

A contract is 'a promise or set of promises between parties, which the law will enforce'. This is a simple working definition, to which we need to add some additional basic features:

- **The parties.** The parties to a contract have to intend to enter into a legal relationship. Typically, the relationship provides that one party will offer to do something beneficial for the other, who undertakes to reward the first party in some way.
- **Agreement.** The contract must be in the form of an offer (i.e. the 'promise') which has been duly accepted. The promise in such a form sets out the obligations of the promisor and the promisee.
- **Consideration.** The promise is supported by a consideration, which is some benefit conferred on the promisor (or detriment suffered by the promisee) such as the contract price.
- **Dates of coming into effect and of completion.** The date of coming into effect is the date when the agreement between the parties formally begins. It may be qualified by certain conditions (*conditions precedent*) which have to be fulfilled before the agreement can begin. In project contracts, the formal end of the agreement is usually referred to specific events, rather than to a calendar date.
- **Proper law.** It is useful – most people would say vital – to state the legal code under which the contract will be interpreted should any dispute arise.

Several other clauses are also important, depending on the circumstances of the project. The main aspects of the contract agreement which should be covered are:

- definitions of the buyer, the seller, the documents which make up the contract, and the contract works;
- scope of supply – usually a list of what the seller undertakes to supply (note the problem of inclusions *versus* exclusions), and usually including spares and consumables related to a defined period of time;
- timing of the programme for carrying out the contract works;
- delivery stipulations as to inspection and testing, marking, effecting delivery, protection and packaging for delivery;
- ownership of goods – defines when ownership of, and responsibility for, the goods passes from the seller to the buyer;
- installation – codes of practice and/or regulations to be observed, stipulations as to carrying out site work, and the issue of materials for installation;
- changes – responsibilities for correcting discrepancies in documentation and for modifications or variations in work;
- liability for defects – defines the responsibility of the parties in the case of defective supply or related work;
- terms of payment – amounts, timing and method of effecting payment;
- confidentiality – responsibilities of the parties for protecting confidential information, especially in connection with intellectual property rights;
- miscellaneous, e.g. extension of schedule, assignment and subletting, force majeure, resolution of disputes, termination, cancellation, seller's default, bankruptcy, liquidation, addresses for notices.

13.3 Typical contracts

Usually, in project contracts, the parties can be considered to be a *seller* (there may be more than one: e.g. contractor, equipment vendor, licensor, etc.) and a *buyer* (e.g. owner, client, licensee, etc.). The promise is to supply, for example, designs, technology, plant, a complete installation etc. The consideration may be money, goods, reciprocal services (as in cross-licensing) etc. There will be other stipulations too, such as confidentiality, the ownership of intellectual property, dealing with events which are outside the control of the parties, dealing with disputes, and so on.

EQUIPMENT SUPPLY

Typical contracts for the supply of equipment contain clauses such as those mentioned above. However, much equipment is offered under seller's 'standard conditions of sale'. Also, many buyers issue 'standard conditions of purchase'. These 'standard' conditions very rarely match, especially when the parties are of different nationality. The United Nations organization has attempted to produce and popularize 'general conditions', but without, it seems, very great success.

ENGINEERING SERVICES

The supply of engineering services has the characteristic that such services have no intrinsic value except for the particular buyer in relation to a particular project. Consider, for example:

1. supply of a design package
2. supply of technical assistance or advice.

A *design package* consists of a collection of drawings, specification sheets, calculations, manuals and similar documentation. While the scope of supply is easy to specify, the quality of information contained in each type of document may be difficult to define. The design package is, in practice, defined by reference to the design basis and operating philosophy for the project, the battery limits (BL) definition and specification of tie-in points, the extent of information to be provided in each document type and the delivery programme.

Technical assistance may be described, but is largely undefinable except in terms of man-days of commitment of specified categories of personnel, e.g. numbers and types of specialists, duration of commitment, timing for each type, rate per man-day and allowable expenses. A technical assistance contract will also contain clauses regulating such things as reasons for withdrawing and replacing specialists, facilitating their travel, providing working and welfare facilities, etc.

Definition, and therefore control, of cost-effectiveness is helped by setting out the limits of the specialists' responsibility (and authority); normal/abnormal work and rates; the working conditions required; and the responsibility of each party for providing them.

Technical assistance, especially that involving training of local personnel, often falls within a technology licensing agreement. It may then be governed by a separate contract the terms of which must be coordinated with the terms of other related contracts.

COMPLETE PLANT TURNKEY SUPPLY

Complete plant turnkey supply will include provisions for both equipment supply (e.g. construction equipment and tools, and special materials such as catalysts) and engineering services (including technology and technical support). The contract philosophy has to be consistent: it is important to avoid the possibility of contradictory positions evolving for different elements of the supply. Keep in mind the following:

- Equipment may be delivered in a completely fabricated form (e.g. pumps), part fabricated for completion on site (e.g. columns) or supplied as plate or pipe for on-site fabrication.
- Bulk items, such as instrumentation, cabling and pipe, and bulk materials, such as structural steel, concrete, paint and insulation, require careful housekeeping arrangements for storage on site before use.
- The first inventory of materials such as column packings, catalysts, absorbents etc. may require special provisions for handling and storage. The first inventory also includes consumables such as lubricating oils and greases, sealants and the like.
- Consumables, spares and spare parts tend to get used at an abnormally high rate during pre-commissioning and start-up.
- Special tools for erection, if forgotten or damaged, can usually not be replaced from local sources at a remote site. Their non-availability will cause delay.
- Technical services will include the various process and functional disciplines, procurement, expediting, inspection, quality assurance, project management, supervision of erection and commissioning. Even with a turnkey contract, the buyer will often assign personnel to take part in these activities. It is important that their participation is regulated by the contract in such ways that the project benefits as a whole, or at least does not suffer, as a result of their participation.

Other technical services will be supplied in which buyer's personnel are not usually involved, in the sense of being assigned to participate in the seller's functions. These include commercial management of shipping, insurance, finance and project accounting.

The basic instruments of definition for all these supplies, which typically form the technical annexes to the contract, are:

- the 'battery limits' definition, plot plan and tie-in points;
- equipment, materials and services supply lists;
- schedules for delivery, construction and commissioning;
- performance guarantees.

STANDARD FORMS

Quite apart from the standard conditions of sale or purchase mentioned above, a number of standard forms of contract have been developed by the professional engineering institutions (in Britain), FIDIC (for consulting engineers operating internationally), the World Bank (for suppliers of goods or services under WB aid programmes) and others. Many large operating companies, such as the oil and chemical majors, use their own standard forms. So, of course, do government departments.

The basic justification for standard forms of contract is simply to facilitate the conduct of trade, especially where the objects of trade are technically and commercially complex, as in engineering projects, while the parties work frequently together and have established a common business 'culture'.

However, we should not assume that standard conditions, even when used to regulate contracts under these fairly ideal conditions, operate to apportion risk and reward fairly and equitably between the parties. On the contrary, they may be used by powerful buyers to oppress relatively weaker sellers (e.g. when highly complex consultancy services are demanded under terms designed for straightforward construction contracts), or by knowledgeable sellers to mislead relatively inexperienced buyers (e.g. in some technology transfer projects for developing countries).

The building and engineering industries make extensive use of standard forms, though often as models on which to base negotiations. An interesting observation by lawyers is that while the standard forms in use in both industries are quite similar, they result in considerable litigation in building, and very little in engineering. The reasons for this are not clear. Suggestions include a greater potential for on-site disputes in building; builders' clients (such as local authorities) are more litigation-minded and consequently more likely to seek a resolution of contractual problems through the courts. It is perhaps also the case that building contracts are more open to misinterpretation: it has been said that the standard form of building contract 'works very well in practice so long as it is not read' (Tillotson, 1985).

From the point of view of the project manager considering how to run a project efficiently under a proposed standard form of contract, probably the most important issue is to identify where the standard form imposes inefficiency. For example, if the standard form requires the project to be run under a lump sum price regime, it should be sufficient for the project manager to provide only *time* auditing information to the client (except for engineering or site changes, which need both time and money expenditure data). This sort of problem arises especially when a change from one price regime (say, 'reimbursible cost + fee') to another (say, 'firm price') is envisaged, as the project moves from an early phase to a later one.

Because of the inherent intricacy of engineering operations, engineering contracts tend to be very much concerned with establishing the parties' rights to be indemnified against losses. This extends to third parties, and notably to the chain – which is often composed of several links – of subcontractors. It is often in this area that a standard form becomes potentially oppressive, as in the case of a powerful buyer seeking to establish the right to recover from the seller any loss, from whatever third parties, arising from performance of the contract.

It seems fair to say, in summary, that the benefits of standard forms of contract accrue to the (corporate) parties, while the concomitant problems accrue to the manager of the project. The project itself is, of course, non-standard. In this connection, the 'New Engineering Contract' (NEC) is an interesting development. Published by the Institution of Civil Engineers (UK), the NEC places considerable emphasis on project management, and strengthens the role of the project manager, at the same time clarifying his responsibilities (Birkby, 1993).

An aim of contract strategy will therefore be to determine to what extent acceptance of standard forms is justified in the context of the particular project being considered.

MANAGEMENT CONTRACTS

Under a management contract, the *manager* (who may be an individual or a firm) assumes responsibility for the management, including operational control of an organization, which remains owned and financed by the owner. Usually, the management contract provides that the manager is paid a fee for the assignment. Sometimes the manager may have separate financial interests in the organization he manages. However, the key feature of a management contract is that the manager is given *authority* to manage, and not just to provide advice or consultancy services.

Management contracts are well known internationally in, for example, hotel operations, health care, transportation and some rather straightforward industries such as cement manufacture. In these, they appear to operate satisfactorily. In more complex industries, such as chemicals, steel and similar process operations, their record is mixed.

Sometimes, project development contracts are brought into being which have some of the characteristics of a management contract. The owner of a complex project retains a specialist (individual or firm) to manage and coordinate design, procurement and/or construction. Often, the specialist subcontracts out these activities to other parties. Under these arrangements the specialist is responsible only for the physical development of the project, not for the management of the business which the project is intended to support. In some cases, however, the specialist may also be made responsible for training local personnel, for commissioning, and even for running the plant, marketing products and generally conducting the business for a certain number of years. In such cases, the arrangement becomes a fully fledged management contract.

Usually, management contracts are adapted to each particular situation, but there are four general factors which are important for success:

- the 'external environment' should be supportive;
- the organization must be fundamentally viable;
- the owner must be enthusiastically behind the management contract concept;
- the manager must be given full authority and control of the organization so as to be able to meet the aims of the contract.

If these factors are missing, the management contract will probably not achieve its aims.

Management contracts are often politically sensitive, and difficult to structure and operate in a volatile situation. Consequently, the 'scope of work' element of the contract should be defined with particular care, especially as regards objectives and performance criteria. The choice of the management team is also very important. A checklist of strategic items to be considered when planning work under a management contract would probably give high priority to the following aspects of building a satisfactory relationship between manager and owner:

- clear definition of each party's responsibilities;
- the scope of the manager's authority and control;
- the facilities, services etc., which are to be provided by the owner;
- processes and procedures for communication, decision making, monitoring and control;
- personnel;
- training and transfer of technology and knowhow;
- sound financing arrangements.

Like all types of contract, there are pros and cons for management contracts too. A management contract will reduce the owner's workload; make use of local resources; allow an early start to the project; and offer potential advantages in purchasing, especially international purchasing. And the project remains under the ultimate control of the owner. But the owner carries most of the project risk and does not know the total cost of the project until after completion. Furthermore, additional levels of management are brought into the project, which may lead to confusion and delay.

13.4 Untypical aspects

While it is normal – typical – to adapt standard forms of contract to suit particular projects, there are some aspects of proposed contractual relationships which, being untypical, are potential pitfalls for the unwary. Three of these are touched on below. They by no means exhaust the range of untypical aspects (or pitfalls), but they are useful examples of what should trigger alarm bells in the mind of the project manager.

LETTERS OF INTENT

It happens quite often that the parties agree to begin work before a formal contract comes into effect. This work may be undertaken under a 'letter of intent' – a document stating that the writer (usually the project sponsor) *intends* to enter into a contract with the addressee (usually a contractor).

Beginning work on the strength of a letter of intent enables the contractor to gain time in what may otherwise be an extremely tight schedule, and he may have other reasons for making an act of faith in the sponsor's intentions. He is evidently put in a position of some vulnerability, for the (English) courts normally consider a letter of intent in the same light as an agreement entered into 'subject to contract', i.e. no binding obligation has been created unless and until the contract comes into effect.

However, the courts have been known to deem such a contract to have come into effect from the date of the letter of intent, that is, to have retrospective effect. Then, even if no real contract is ever entered into, it may be possible for a contractor to recover some payment for work done and accepted under a letter of intent. In such a case, the court will decide what is reasonable, on the basis of *'quantum meruit'*. The court's view is unlikely to match the contractor's expectations.

The situation arising from a letter of intent is an example of the grey area which exists between contract law, with its apparently clear-cut rules governing the working relationships, and actual practice where the pressures of business and the marketplace make it impossible – or impractical – always to hold to the strict letter of the law. If a dispute does arise, the courts are likely to try to infer practice from precedents of regular patterns of business dealings in the past. However, even where such precedents exist, there appear to have been very many alternative interpretations of the obligations created by letters of intent. Where there is little previous experience to guide a court, the outcome of a dispute is likely to be no different from a lottery.

LETTERS OF COMFORT

A rather similar situation arises from 'letters of comfort'. These purport to confirm the policy of a parent company to support fully a subsidiary, for instance, in the case where the subsidiary is a party with others to 'joint and several' liability in respect of obligations undertaken in a contract with another party. The latter is thereby 'comforted' that, should the liabilities fall due to be settled by the subsidiary, the resources of the parent will be available to support the subsidiary. The (English) High Court has, however, declared that a letter of comfort is not equivalent to a guarantee (in connection with joint and several liability for the costs of abandoning an oil rig).

COLLATERAL WARRANTIES

The intention of a collateral warranty is to pass the obligation (and cost) of rectifying defects to another party, who may have no direct contractual relationship with the parties to the project contract (or, sometimes, who is already in a contractual relationship which is, however, extended or modified by the collateral warranty). A typical application is to give to a developer or financier (or even the tenant of a building) the right to recover losses directly from any party alleged to be responsible for a defect causing them loss, such as the designer, constructor or project manager.

In some cases, acceptance of collateral warranties is made a condition of the appointment of such (professional) parties. Since the warranties are usually drafted by lawyers retained by, for example, the developers, they are drawn to cover all the possibilities of loss against all likely parties. And because collateral warranties are considered likely to affect others to whom the obligations may be assigned (thereby creating an apparently infinite scope for claims), professional indemnity insurance is difficult – and therefore expensive – when such warranties are involved.

Collateral warranties are a fairly recent phenomenon in the UK, and their impact in practice has not really been tested in the courts. In the EU, however, single project insurance obviates the need for collateral warranties.

13.5 Contracts and project risk

The first question that arises is: 'Why have a contract at all?' Usually, at least to begin with, if we are owners, we are thinking about using our own in-house resources, rather than an outside organization such as a contractor. In fact, even if we do use in-house resources, a relationship similar to a contract will have to be established between 'us' – as buyer – and the in-house resource – as seller – if the work is to be properly controlled. Indeed in many organizations, internal relationships of this kind are embodied in a formal 'contract' document.

The main reasons for establishing a contract are, however, less to do with formality than with factors such as the following:

- we may not have sufficient or suitable resources to undertake the work ourselves;
- we may have the resources but be unable to make available, at the right time, the right skills and knowhow;
- we want to reduce the commitment of our assets or resources to the project, or to some parts of it;
- our project may be in a type of business which is new to us, so we prefer some more experienced party to be responsible for the project;
- we want to spread the risks and are prepared to accept less direct control as a consequence.

All these confirm the characteristics of a contract as an instrument for *balancing* risks and motivations. I stress the word 'balancing', for both parties should carry some of the risks in a project. If a contract transferred all the risks in a project from party A to party B, only B will be motivated to manage the risks so as to achieve project success.

The risks are, in brief, that the project will fail to come up to the parties' expectations. At the level of the basic criteria for project success (or failure), these are expectations of completing the project within budget, within schedule and in accordance with its performance or duty specifications. Although the parties to the contract – the buyer/owner and seller/contractor – have different perspectives on these expectations, their viewpoints on the criteria are similar: failure imposes some kind of penalty. For example, for the owner, failure to complete on schedule potentially entails loss of revenue. For the contractor, it entails the continued cost of committing resources which should be available for work elsewhere. In either case, the affected party suffers a penalty in relation to its expectations.

Clearly, ingenious people will seek to find ways of avoiding penalties, and other ingenious people will seek to find ways of frustrating the first group – hence the tortuous complexity of penalty clauses in many contracts. In the context of contract strategy, however, we should recognize that:

- all projects entail risks which affect, to a greater or less extent, both parties to a contract;
- nonetheless, both parties intend that the project will succeed, and both should be motivated to achieve that end;
- if operational losses are incurred, they are not likely to be offset by any feasible penalty payment;
- while a penalty may act like a stick to drive the parties on, their motivations are just as likely (perhaps, more likely) to be strengthened by incentives, which act like the carrot in the proverb.

There are not many basic contract types. They can be reduced to three, although there are many possible permutations and adaptations which utilize different aspects of the basic types. The basic types are:

- Lump sum
- Schedule of rates/bill of quantities
- Reimbursible.

Each type effects a different balance between the risk accepted by the buyer (or owner) and the information available or needed to control the project (Fig. 13.2). In essence, a lump sum type of contract minimizes the owner's risk, but requires a lot of information in order fully to define the work. A reimbursible contract type requires little information to define the scope of work, but exposes the owner to considerable risk. It will be obvious that situations arise where this is unavoidable, but they should not be allowed to continue for very long. Converting from a reimbursible to a lump sum contract regime is therefore quite a common practice.

Recently, attempts have been made to set out contract conditions which encourage cooperation between owner and contractor. An example for this is the 'preferred contractor' agreements between some petroleum industry majors and selected contractors who have worked frequently for them, and so have become familiar with the majors' project work practices. The aim of these agreements is to establish 'best practice' as a norm and to avoid, as far as is practicable, divisive or protective clauses, in the interest of speeding up schedules and saving costs, especially in projects involving new technology where teamwork is recognized as the best way to solve problems associated with 'concurrency'. One model is the National Economic Development Council's report *Partnering: contracting without conflict*. Partnering contracts have been

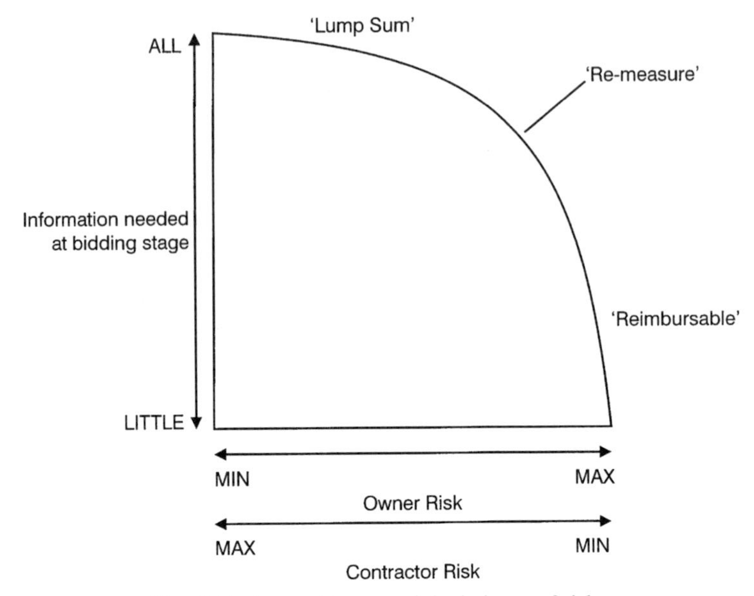

Fig. 13.2 Contract types and the balance of risk.

extolled as having led to significant cost savings. A detailed description of how this was achieved in one difficult project has been published recently (Knott, 1996).

Generally speaking, if the work content and scope of the project can be clearly defined – and this definition is accepted by both parties – it makes sense to use a lump sum type of contract because this at once motivates the contractor and minimizes the owner's commitment of resources. If the work content and scope are not clearly definable, the owner will have to assume substantial control of the project, and it is preferable to use a reimbursible type of contract.

Sometimes, both parties can agree that initial work should be carried out on a reimbursible basis until the project is properly defined, when the contract regime should be changed (following agreed procedures) to, say, lump sum, possibly adopting a 'work package' approach as the work is progressively defined.

A frequent bone of contention is whether, granted these pros and cons, a lump sum contract actually works out cheaper than a reimbursible one. Owners tend to take the view that lump sum is normally bound to be cheaper because a reimbursible contract is tantamount to paying the contractor merely to attend, not necessarily to do any real work. Contractors point out that it is they who carry most of the risk in a lump sum contract, so they are entitled to build a risk premium into their lump sum price. Cynical owners reply that contractors will naturally promote the myth of expensive lump sum contracts so as to encourage owners to agree to reimbursible arrangements, thereby allowing the contractors to rest easy and retire in luxury. Clearly the choice of one regime or the other must take account of the business 'culture' of both sides, as well as the nature of the project.

Owner companies frequently seek to reduce contractual risk by requiring the contractor to post a *performance bond*. The bond provides for the owner to be paid damages if the contractor defaults. But they seem to raise more difficulties than they solve (except for the lawyers). The wording of the bond is crucial. On-demand bonds worry contractors, because ill-intentioned owners may call the bond without reasonable justification and because the issuing bank is likely to reduce the contractor's overdraft facility by an amount equal to the value of the bond. Other forms of bond are available from insurers. These are, unsurprisingly, weighted in favour of the insured contractor, which does not please the owner company. Specialist legal advice is essential if performance bonding is to be part of contract risk management.

There are many ways of incorporating both sticks and carrots in the type(s) of contract deployed. And while the extent of direct control exercised by the owner will vary, the owner must always assure himself that the overall quality standards he has set are met, even though the contractor will be responsible for the quality of his own work. This will inevitably be reflected in the selection of contractors, which is an integral part of the contract strategy.

An outline of the reasons for and against the basic contract types is given in Section 13.7, which also contains notes on variations, penalties and incentives.

13.6 Developing contract strategy

CONTRACT STRATEGY AND LEADERSHIP

If we suppose that each project is unique, a project contract represents a one-off trading encounter between the contracting parties. That is to say, it is a representation of trust in terms of each party's expectation that the other will be honest or will cheat. We can visualize a limited number of 'game' situations:

- I will if you will; I won't if you don't (*'Assurance'*).
- I will if you don't; I won't if you do (*'Chicken'*).
- The *'Prisoner's dilemma'*, in which each party expects to gain by cheating, whether the other party is honest or not. In this case, if you (the other party) are honest, it pays me to take your goods and refuse to pay. Or, if you cheat, it pays me not to honour my obligations in the first place. In this case – given a one-off encounter – honesty is never the best policy.

Consider a price negotiation taking place with a view to signing a contract. Each party hopes the other will take a soft line so that he can win a concession by being hard. But both parties would rather concede than let the agreement fall through because they both believe that mutual concession is mutually advantageous compared to no trade at all. Hence:

- both parties prefer to be honest whether the other party cheats or not (*'Harmony'*).

If we consider only material incentives to trade, 'Harmony' is rare. Assuming that the characteristic feature of leadership is the capability of manipulating trust, then the role of project leader offers the possibility of converting 'Assurance', 'Chicken' and 'Prisoner's dilemma' situations into 'Harmony'. How? By considering that people respond to moral as well as to material incentives. By associating guilt with the intention to cheat, the leader has the possibility of inducing honesty whether he expects the other party to be honest or not. Provided that the feeling of guilt is sufficiently strong, it will always pay to be honest: the result is 'Harmony' (Casson, 1991).

But if we consider that a project contract is not, in fact, a one-off encounter, but simply one example of several similar encounters, we may come to learn that other parties are dishonest. Then, supposing that the moral manipulation, and hence our feelings of guilt, are only moderately intense, the encounter system will tend towards an equilibrium of mutual cheating! (The most successful gaming strategy in this situation seems to be *'Tit-for-tat'*, in which you are always honest in the first encounter, but after that you do exactly what the other party did last time (Axelrod, 1990)).

In this connection, it is worth noting that the real benefits that can follow from cooperation can only be obtained by long-term commitment, which implies trust. If our overriding aim is to maximize profits by all means, commitment and trust will be discouraged. Moreover, motivation by penalty provisions depends on the discount rate applied to future profits: a high discount rate encourages cheating, for the future loss represented by the penalty will appear to be small. We do not only have a commercial issue to resolve, but a moral one as well (Hutton, 1995).

To the extent that moral manipulation makes businessmen feel uncomfortable, we have an alternative: enforcing agreements by law, i.e. through a contract. Which is likely to be the more cost-effective? Well, reliance on contracts can be expected to be cost-effective if:

- the system of contract law is highly developed;
- there is little doubt about what constitutes the quality of performance;
- we are working in a low-trust culture;
- the available leadership is ineffective.

CONTRACT STRATEGY AND PLANNING

After the philosophy, we have to be practical. From our previous observations, we can list a few fundamental issues to consider:

- **The scope of the work.** Is the project going to use established technology, or is it at the limits of technological knowledge? Does it depend on new, possibly unproven technology? Is the design conventional, or does it rely on new concepts? Can the work scope be clearly defined, or does work have to be done to define it? Does the design itself, in relation to work scope, influence the structure of contracts (e.g. modular designs may influence construction; prefabricated items influence subcontracting; very large items may influence the way in which site operations are managed). Note that these considerations may be affected by differences in the 'culture' of various industries.
- **Priorities and objectives.** Are these clearly defined? Is the project schedule likely to be accelerated? If so, what incentives can be built into the contracts? Is it feasible to run separate contracts in parallel, or to combine some activities under a single contract? How will productivity be measured?
- **In-house resources.** Is your organization able to commit in-house resources to the project, or to certain parts of the project? What assumptions have been made about the availability of specialists (especially in activities involving innovation)? Are any additional, temporary resources available? What additional training may be needed?
- **External relations.** Are relevant contractors and suppliers likely to be hungry for work, or too busy elsewhere to be very interested? What possibilities exist for reconsidering rejected contractors if the preferred ones prove to be uncompetitive? If there are parts of the project which involve long lead times, will it be necessary to consider ordering early or reserving design or manufacturing capacity? Are there any features of the contract which will be specially interesting to the government, joint venture partners or the community at large? How can these be mobilized to benefit the project?
- **Contract price terms.** There are two basic options:
  1. negotiate payment based on the 'earliest completion dates' in the project schedule;
  2. negotiate payment based on the 'latest completion' dates in the project schedule.

  The choice affects the planned S-curve for the project. Compromise payment schedules can be related to 'level of effort' or other progress schedules (see Chapter 12.2).

These considerations give rise to the elements of the contracting plan:

- It should be developed early, as an integral part of planning in development of the project.
- It should take account of the project backcloth, and the influence these 'outside' factors exert on the overall project strategy.
- In particular, it should set out the way the project will be divided into contract work packages, the interfaces between these, their relation to sources of supply, and the general method of managing contracts.
- It should consider alternatives, evaluating each in terms of resources, schedules, budgets, finance and risk.
- It should state preferences for contract types and their aims, together with fall-back positions, in order to achieve the project objectives. It may be appropriate to use different types of contract for different parts, or phases, of the project.
- It should indicate the required management organization for the types of contract foreseen, and propose a short-list of preferred contractors, thus establishing the initial guidelines for negotiation.

13.7 Basic contract features

LUMP SUM CONTRACTS

There is a firm* price for the work. Provisions for contract price adjustment due to increases in labour and materials costs can be included. Changes should, if possible, be priced as a lump sum.

Advantages

- Contract cost is known at the outset.
- Contractor assumes greatest risk; owner assumes least risk.
- Strongest motivation of the contractor due to his commitment on time and cost.
- The high degree of contractor responsibility means simpler contract control by the owner.
- More opportunity to select the best contractor because of firm price quotations and precise planning.
- Usually lower investment cost.
- Contractor will tend to use his best staff on this type of contract because performance has a large effect on his profit.

Requirements

- Precise definition of the work scope.
- Reasonable financial risk to the contractor.
- Stable market conditions and absence of major economic or political uncertainty.
- Close control of changes.

Disadvantages

- Contract award may be delayed later than with other types because of the need to define work fully, so that the tender period will be longer.

* Note: a *firm* price allows for escalation, usually according to an agreed formula. A *fixed* price does not.

- Owner's ability to influence performance within the terms of the contract, especially as regards cost, is reduced because of the extent of the contractor's responsibility.

Owner's manpower commitment

- Owner's manpower commitment is at a minimum. Contractor exercises control of the work. The owner's role is essentially to monitor the contractor's performance, but the owner must also ensure quality because the extent of the contractor's responsibility carries the risk that quality may suffer under pressure to maximize or retain profit.
- Owner must closely monitor changes through clearly established procedures. The effect of proposed changes on time and cost must be evaluated throughout the whole programme.

REIMBURSIBLE COST

This includes the contractor's costs for the work, plus a profit. The profit component may be a fixed fee, a rate related to time, or a percentage of the cost. Note that there is a slightly stronger motivation of the contractor on a cost plus fixed fee type than on a simple cost plus contract.

Advantages

- Owner's ability to influence performance is greatest.
- Contract award may be earlier than with other types because the work scope does not have to be well defined, so the tender period is shorter.
- Change control is simpler.

Requirements

- Due to the high risk assumed by the owner, and lack of clear definition of the work scope, requirements, e.g. for project information, are minimal.
- Where the add-on element is a fixed fee, the work must be defined sufficiently for the contractor to be able to estimate the extent of the services he is likely to supply.

Disadvantages

- Cost is unknown until the end of the contract.
- Owner retains greatest risk.
- Contractor is not required to apply direct control himself.
- Identification of the best contractor may be difficult.
- Control of work by the owner demands considerable effort.

Owner's manpower commitment

- Owner's manpower commitment is at a maximum. Owner must assume active control and direction of the work. This requires an appropriate organization to cover all control elements.
- Contractor is essentially responsible for technical supervision, labour and labour relations.

SCHEDULE OF RATES

This is a series of firm or fixed prices or rates for units of work output set forth as an itemized list. Quantities are either not given against the schedule of rates or are only approximate. Note that the *bill of quantities* type is similar to the schedule of rates, but quantities are accurately specified and costed on a unit price basis.

Advantages

- Similar to lump sum, but the owner assumes risks related to determining the quantities. The advantages of lump sum are therefore present, though to a lesser degree. The contract cost is not clearly defined as in lump sum.
- Change control can be somewhat simpler than lump sum, as changes in quantities can be picked up by direct measurement.

Requirement

- Similar to lump sum, but accurate definition of the extent of work is not necessary. However, where no quantities are given, the contractor needs some indication of the scope of the work, because this may affect his rates.
- All work must be covered by the rates quoted.

Disadvantages

- Similar to those for lump sum, but greater control by the owner is called for, as there is not such complete definition of the work scope. In the case of a bill of quantities, the definition of the material work scope is closer to that needed for lump sum.
- Final cost is not well defined.
- This type is more suitable for contracts such as fabrication, erection and pipework than for plant supply and installation.

TIME RATE OR DAYWORK

This is a series of time-based fixed prices for units of work input.

Advantages

Similar to the reimbursible type, but as the contractor quotes fixed rates he assumes risks related to labour compensation and equipment operating costs.

Requirements

Precise definition of work scope is not required.

Disadvantages

Similar to reimbursible contracts, with the owner taking risks related to effective performance or productivity.

Owner's manpower commitment

Similar to reimbursible contracts.

WORK PACKAGE

There is a fixed price for the work in each work package. The initial work package constitutes the basic contract to which successive packages are attached.

Advantages

These are similar to lump sum contracts, but as definition of all the work is not required at the outset, contract award may be advanced so as to speed up project completion.

Requirements

Similar to lump sum contracts. The first work package must contain the basis for pricing and scheduling the later, negotiated packages. Each work package must be discrete, with only minimal and well defined effects on other work packages.

Disadvantages

- Similar to lump sum, but greater owner's commitment is required to control interfaces between work packages.
- Successive work packages would in principle be negotiated with the original contractor rather than put out to competitive tender.
- Where work packages are let to different contractors they become separate contracts.

CONTRACTUAL PENALTIES, LIQUIDATED DAMAGES AND INCENTIVES

A *penalty* is a detriment, usually a sum of money, inflicted on one party by the other for failure to perform his contractual obligations as to the agreed quality, schedule or cost. *Liquidated damages* are in essence a genuine agreed pre-estimate of damage suffered by one party because of the other's failure to perform. They have the characteristic that there is no intention that they should be paid if the contract is properly fulfilled. In practice, liquidated damages are often set as the maximum amount payable by the defaulter under all circumstances of failure to meet the terms of the contract.

The real purpose of penalties – and liquidated damages – is to motivate the seller to perform his obligations, even though this costs him more than he originally estimated, because this will be less expensive to him than paying the penalty. However, a seller may find himself in a situation where paying the penalty seems to be his best way out. He is then likely to lose all motivation to do his work properly, and so we need to take care that the penalty provisions bear a reasonable relation to the allocation of project risk. (In some countries, the courts may consider the amount of damages stipulated in the contract as the minimum rather than the maximum damages recoverable.)

Penalties are not, in practice, very good motivators of performance. This function can be improved by, for example, *sanctions* (withholding payments; but this provision needs to be very carefully drafted); *expedition* (buyer's right to expedite work at the expense of the seller); or suitably designed incentives.

Incentives are used in contracts to motivate contractors towards reducing the time or the cost of the work. Like penalties, they are features added to other contract types, rather than being a separate type. Incentives relate either to the magnitude of payments made to the contractor or to their timing. The latter is usually the more effective.

Advantages

There is higher motivation of the contractor either because improved achievement means more profit or because payment is linked to completion of physical stages of the work.

Requirements

The different types of incentive arrangements have different requirements. A common form is to set the duration of the contract and to make predetermined payments for early completion. The incentive is more money for an advanced end date. Such arrangements should be included in the main contract. This form applies to lump sum and related types of contract. Milestone payment terms may be used as an incentive. Milestones are set for physical completion of stages or sections of the work, and payment is linked to the achievement of each milestone. This only affects the timing of payments and does not involve extra money.

Target cost incentives can be used with reimbursible or related types of contract. A target cost is set, and any savings or losses on the target are split in a predetermined proportion between the contractor and the owner.

Disadvantages

Although incentives are intended to confer specific advantages, one disadvantage may be the potentially disruptive effect of changes on the incentive provisions. Incentives may also have an unsettling effect on the labour force, if workers feel they are not sharing in the benefits.

Owner's manpower commitments

The use of incentives usually means additional effort from the owner as regards change control. The effect of changes on incentive provisions has to be determined and suitable adjustments made to ensure that the incentives are not weakened. Some additional effort will also be needed to calculate the incentive payments to the contractor.

13.8 Contract control

Methods of control and progress reporting vary in detail according to the size of the project and the type of contract. However, the basic requirements for contract control remain the same:

- Communications with the contractor are to be channelled through one named focal point in the project team.
- Progress reporting mechanisms are written into the contract.
- Change control mechanisms are written into the contract.

All contracts should contain the following control procedures:

- a formal project implementation plan, e.g. a precedence network and project activities lists;
- agreed milestones and deliverables;
- coordination and administration with the appropriate rights of access and audit;
- resource schedules;
- types, layout and frequency of reports;
- requirements for additional data;
- control of safety and quality;
- change control and claims;
- close-out.

Contract control of site work requires that the owner's project site team be set up so that it is appropriate to the type of contract, the capability of the contractor and the complexity of the project. The project site representative must be given adequate authority, and his authority must be clearly explained to the contractor. Site administration facilities, e.g. accommodation, offices, secretarial services, bank account and transport, should be put in place as early as possible. Contractual obligations and requirements should be observed from the outset.

Problem areas for contract control are:

- contract variations
- invoicing and payments
- claims
- close-out.

They should be carefully considered in the contract plan so that effective procedures for managing them can be incorporated into the contract documentation.

13.9 Summary

A contract makes formal the arrangements between a project buyer (e.g. owner) and seller (e.g. contractor) to regulate their respective commitments to the project. The contract is the outcome of negotiation. It has the character of a one-off agreement for a one-off working relationship. As such, it is important to consider carefully the approach to planning the contract.

Although specific to a particular project, the contract will contain several typical elements. Standard forms of contract are available as models, which are useful as the bases for detailed negotiation. However, untypical aspects may also appear, often as a result of one party's attempts to bias the agreement in his favour, and so distort the function of the contract as a device for allocating project risk fairly between the parties.

While there are not many basic types of contract, there are many possible permutations. Choosing the most suitable type depends on the nature of the project and the resources available (especially to the owner, as buyer). In making the choice, we need to look for a contract type and contract conditions which will encourage cooperation while ensuring that project work can be properly controlled.

Different types of contract offer different advantages, which carry with them different disadvantages. In a large project it may be best to use different types of contract for different parts of the project, making sure, however, that the different contract types are consistent with one another, otherwise overall control will be jeopardized.

14
Aspects of Quality Management in Projects

14.1 Introduction

We take care to define our project by setting boundaries around what we, as project people, are responsible for. We specify 'on-site' and 'off-site' facilities; we define the 'battery limits'; we make lists of what is included and what is excluded. All this tends to make us – and others – feel that the project is a special kind of fiefdom on its own, as indeed in some ways it is.

However, no project exists in isolation from the rest of the world. It impacts on other people (the community at large as well as the people working on the project) and on its environment. The impact of the project on the world outside it is well expressed by the concept of 'technical integrity', i.e. the condition met by the project when it represents no foreseeable risk of failure endangering the safety of personnel, environment or asset value (Section 5.1). Technical integrity brings together aspects of quality management and environmental management. However, it treats 'quality' in a special way.

14.2 Quality

First of all, consider this definition of 'quality': 'the total of features and characteristics of a product or service that bears on its ability to satisfy stated or implied needs' (ISO 8402). It is interesting to compare this definition with 'technical integrity'. The concept of technical integrity is *risk-oriented* (and we may note in passing that the 'foreseeable risk of failure' can never be zero!). The ISO definition is distinctly *market-oriented*, because it is the market (in an entrepreneurial system) which establishes the satisfaction of 'stated or implied needs'. We should keep these somewhat different orientations in mind when we consider project-related questions of quality management.

DEFINITIONS

Quality assurance (QA) is concerned with all the activities that affect the quality of a project. QA integrates quality planning (QP), quality control (QC), quality improvement (QI) and quality management (QM).

QP is concerned with setting project quality goals and strategies, and the means of converting these into definite objectives, policies, procedures, standards and working practices, and devising suitable controls. QC covers the application of controls, inspection and testing to detect deviations and to determine the actions needed to prevent a recurrence. QI is the search for more effective ways to manage quality. QM is concerned with the implementation of all these activities.

The project leader will therefore be concerned to ensure that QA and QI principles are built into project objectives from the outset, because it is extremely difficult to put them in place after the project has begun. For the project manager, 'quality' primarily consists of achieving conformity with the requirements of defined project objectives. The emphasis is on conscious planning to determine and control project activities in conformity with these objectives.

THE PROJECT QUALITY MANAGER

Everyone engaged in a project must be concerned about, and involved in, QM. Some people may have specific responsibilities (e.g. for quality in design, fabrication, construction etc.).

Throughout the project it is necessary to monitor and audit QM activities so as to provide documented evidence that all specified quality requirements have been fully met. This applies to all projects irrespective of their size. On large projects, the work is done by a 'quality manager' who is a member of the project team. On smaller projects, a 'quality engineer' or a nominated project representative will be responsible. On the smallest projects, the project manager may have to do the job himself.

THE QUALITY MANUAL

Let us suppose we are dealing with a project big enough to justify a designated quality manager. The principles which we will discuss here will, however, apply to whoever is wearing the QM hat.

At the start of a project, the project's quality manager prepares a 'quality manual' describing the quality system for that particular project. The quality manual sets out the 'quality plan': all the procedures to be used, audit activities and reporting relationships for all phases of the project. The quality manual is typically based on the ISO 9000 series of international standards.

The quality manual is modelled on corporate quality policy statements, but deals specifically with the particular project in hand. In its simplest form, it contains:

- the project QM policy – project policy statement; project quality objectives; and procedures for using, maintaining and updating the manual;
- organization and resources – project organization with special reference to QM and all related interfaces; resourcing for QM functions;
- scope of the quality manual – schedule of quality-related activities; network showing logic and sequence of quality activities; schedule and description of all audits and reviews to be performed; and description of project quality activities, budgets and deliverables;
- procedures and work instructions – index of procedures, guides and standards that apply to the particular project (note that the quality manager does not determine the working procedures and practices for engineering disciplines; this is the responsibility of the respective discipline heads).

It often happens that all the necessary project procedures are not ready when the project is launched. The first issue of the quality manual should, in that event, contain *as a minimum* schedules of the documents yet to be produced, their contents, the individuals responsible for producing them, the time-scales for completion and the manhours required.

The aim of the quality manual should be to convey, as concisely as is consistent with clarity, for all quality-related activities:

- the objective of the activity, i.e. *why* the activity is necessary;
- the basis of the activity, i.e. *what* has to be done;
- the method, the parameters by which achievement is to be measured, and the required level of confidence in the results, i.e. *how* it is to be done;
- allocation of responsibility to a project team member, i.e. *who* is responsible for getting it done;
- the plan for starting and completing the activity, i.e. *when* it is to be done.

THE PROJECT MANAGER AND QUALITY MANAGEMENT

The project manager must ensure that the quality function faces no impediment in the project environment. Typically, the quality manager reports direct to the project manager, with parallel direct reporting to the corporate quality manager. If, however, dual reporting lines are found to be unworkable, then the quality manager – who must be able to exercise his/her objectivity and independence – should report directly to the project manager in order to minimize conflicts of interest.

QM specialist personnel (or other qualified members of the project team) will be responsible for the development and supervision of quality programmes and, when appropriate, for obtaining certification. Sometimes they will also be responsible for such activities as document control and inspection.

All formal quality activities should be carried out according to written procedures based on proven techniques, and evidence of compliance must be recorded. Careful attention to, and use of, the procedures is reinforced if they are written by their eventual users. The QM specialists can provide functional advice, and a professional editor can help with ensuring clarity and readability.

Quality principles and objectives must apply across all project interfaces. The project manager must ensure that the appropriate level of QM is maintained throughout all phases of his project, and that all the players in the project – vendors, suppliers, contractors, consultants, etc., as well as the project team – are subject to the project quality policy and procedures.

Project control plan

The project control plan, part of the quality manual, serves as an overall control for both the project manager and corporate management. It describes the organization, defines functional roles and responsibilities, and provides schedules for, and specifies participation in, reviews. It should be prepared by the project manager at the start of the project, and should be approved by the appropriate level of corporate management.

While major projects demand to be resourced with specific QM expertise, it often happens that, on small projects, or in a multi-project environment, less experienced people have to make important QM decisions. More reliance must then be placed on the use of standard procedures. Pressure of unfamiliar work, conflicting priorities and the absence of third parties whose cooperation is needed can result in these standard procedures being ignored or side-stepped. The project control plan helps to overcome these problems by ensuring that the review cycle is properly planned, has the commitment of third parties, and provides a documented audit trail.

QUALITY MANAGEMENT IN THE CONTROL OF DESIGN

Before design work is started, quality objectives should be developed in relation to:

- statutory requirements
- performance requirements
- production availability
- safety
- operability
- maintainability
- project life expectancy
- capability for extension
- compatibility with existing facilities.

Quality management in design focuses on safe, reliable operation and maintenance under all foreseeable conditions. Several formal techniques are available, such as failure mode and effect analysis (FMEA), fault tree and event tree analysis (FTA, ETA), and hazard and operability study (HAZOP). They are useful for quantifying the consequences of quality failure and for setting QM priorities.

Specifications

Wherever practicable, the equipment and materials specified should be of normal commercial quality, complying with corporate standards. If the vendor's standards differ from corporate standards, a careful assessment should be made before accepting them. Quality requirements in equipment and materials specifications include:

- functional specification of performance, operational and maintenance requirements
- environmental conditions
- mandatory codes and standards
- essential additional QC requirements
- documentation
- vendor's submission of a quality plan.

Design interfaces

Design control requires that each specification is complete and has been verified; that it includes the necessary quality requirements; and that it complies with other relevant specifications and is compatible with interfacing specifications.

Managing design interfaces is one of the most demanding jobs of the project manager. Important interfaces exist between, for example, disciplines, contractors, structures, functional modules, operations, client/user organizations, the project sponsor and project management, financial organizations, and projects. It is therefore very important to establish interface controls.

Key actions are to define work scopes clearly; to use common design specifications; to document the major interfaces, using interface control drawings where possible for clarity; and to nominate the individuals responsible for each work package.

Design changes

Proposed design changes to baseline specifications must be carefully reviewed before approval to proceed is given. The review should establish all resultant effects on other design elements and disciplines.

Proposed changes which would modify a basic design concept must be referred to corporate management for authorization, because they may have the effect of altering the fundamental project objectives. Changes proposed for the purpose of 'fast-tracking' should be examined with especial care, because they almost always alter the project objectives on which all previous planning has been based.

Design reviews

Design reviews must be carried out by competent engineers in order to check discipline and interdiscipline activities; ensure that progress is being made as planned; verify completed work and identify problem areas.

Design contractors

Potential design contractors should be assessed before bids for design work are invited. The appraisal procedure should include assessment of the contractor's relevant experience, track record, quality system and safety record. A quality plan should be an integral part of the contract.

QUALITY MANAGEMENT IN PROCUREMENT

Vendors

Invitations to bid (ITBs) for the supply of equipment and materials should be limited to those vendors who have previously demonstrated their capability regarding quality systems as well as their technical knowhow, production capability and capacity, financial status and safety record. ITBs and purchase orders should be complete. Quality requirements commensurate with the criticality of the items should be incorporated into the purchase orders.

Criticality rating and certification

Equipment items, materials and sometimes services such as site fabrication are graded by a 'criticality rating'. The rating is based on the hazardous nature of the duty, statutory requirements, availability of standby, or past experience. The criticality rating determines the level of inspection required. In effect, it specifies the quality programme, i.e. the documented set of requirements from which vendors and other suppliers can prepare their quality plans.

Quality plans set out the specific quality practices to be followed in the manufacture of items or the provision of services, including quality control and inspection requirements, certification and procedures for rectifying defects. The project manager is ultimately responsible for approving vendors' quality plans.

Vendors are responsible for the quality of their products, and their quality systems should comply with ISO 9000 or equivalent. Monitoring vendor performance may be effected by, for example, a periodic evaluation of the vendor's quality system; surveillance by resident or visiting inspectors; witnessing inspection and tests; review of QC documentation; and audit of QM procedures.

The project manager is also responsible for ensuring that all necessary certification is obtained in the proper form. In even small projects, this may mean that very considerable confirmatory documentation is involved, with a correspondingly heavy workload. It is easy to underestimate this: adequate staffing must be allocated from the beginning of the project.

Inspection contractors

Inspection contractors may be employed to verify independently the quality of vendor supplies. Before being appointed, it should be established that the inspection contractor has the necessary competence, resources and integrity for the project. Individual inspectors put forward by the contractor should be interviewed and tested.

QUALITY MANAGEMENT IN CONSTRUCTION

Contract strategy is the basis of QM in construction. Contract strategy should provide, wherever possible, that the job specification includes quality requirements; that the tender list is restricted to formally approved contractors who must submit quality plans for approval; that a quality plan approved by corporate management is included in the contract; and that contractor performance is monitored and evaluated. If it is impracticable to qualify all contractors in this way, then each tender should be weighted according to an assessment of the additional surveillance judged necessary to bring the contractor's QM up to the required standard.

Control of materials during construction

The project manager must ensure that material received at the construction site is properly identified and documented, and undamaged. Stored material must be adequately protected. Material released for construction must be as specified and in good condition. Procedures for keeping material and equipment in storage must be designed to preserve the identity and quality of the items.

Fig. 14.1 BP exploration – Miller Project. Corner node on jacket, Nigg Bay (March 1991). Photograph courtesy of British Petroleum.

QUALITY MANAGEMENT IN COMMISSIONING

QM activities during the commissioning stage are intended to verify that the facility has been constructed as designed, meets its performance specification, is safe and reliable in operation, and does not violate design limits.

When commissioning is complete and the facility has been accepted, responsibility for the installation passes from the project team to the operator organization. Project QM activities are then concerned with checking the completeness of as-built drawings and specifications; the availability of vendor documentation for all equipment; and the validity of operating and maintenance procedures in relation to experience gained during commissioning.

QUALITY MANAGEMENT REVIEW AND AUDIT

It is always necessary to check that QM is working as intended. An audit schedule of all quality activities should be included in the project QM programme. In effect, quality reviews and audits give a 'go/no go' decision on whether to proceed to the next stage of work. Timing of the reviews should be chosen to facilitate this decision. However, under time pressure, there is a tendency to relax the rigour of the audit and to plan to carry it out thoroughly at the end of the project. This is dangerous, for if the audit then does show the need to carry out corrective work, probably a great deal of work done in the meanwhile will have to be unravelled and done again!

QM also provides an audit trail. The QA programme should include a schedule of audits of all quality activities. The findings of quality audits include programmes of corrective actions which are reported and agreed, i.e.:

- remedial actions needed to correct any observed deficiency;
- investigations to determine the frequency of deficiencies;
- if a deficiency is an isolated case, whether it is typical of a more general problem;
- if the problem is generic, what improvements can be made to the quality system.

USING ISO 9000 IN PROJECT CONTRACTS

ISO 9000 is a series of standards, first issued by the International Standards Organization in 1987. The series is widely used as the basic guideline for project-based quality management.

As mentioned above, appraisal of contractors includes an assessment of the contractor's own quality management system. Typically, contractors are asked to furnish two documents: (1) a 'quality programme' which describes the contractor's general policies on quality management, and (2) a project quality plan setting out the specific procedures, practices and resources proposed for the project under consideration. A list of quality management issues which should be addressed in these documents is given in ISO 9001 (1987, Section 4). Agreement on the contractor's quality management proposals must be reached before the contract is awarded.

It is not unknown for quality standards to be made obsolete by technical developments, raising difficult issues for QA in technically advanced projects.

14.3 Quality and the control of purchasing

INTRODUCTION

Business organizations exist in order to make a profit for their owners. Let us say that profit is the difference between the 'price obtained for sales (of the product)' and the 'cost of buying in (the necessary raw materials and services)'. In a free market, the sales price is established by the competition. So we can maximize our profit only by minimizing our costs of bringing product to the market. Consequently, controlling the purchasing (sometimes called 'procurement') activity assumes great importance in any business. In addition, the cost of purchasing equipment and materials is, in many industries, the largest part of total project costs.

Control of purchasing is a responsibility of the project manager. Moreover, because purchasing is such a significant part of project expenditures, and purchasing expenditures are largely determined quite early in the development of the project (typically, when the project technology is selected), responsibility for the control of purchasing begins early and continues throughout most of the project implementation phase. Purchasing therefore merits careful attention.

ORGANIZATION OF PURCHASING ACTIVITIES

The purchasing function in a company carries out several roles:

- **policy making** in respect of the market and in response to the company's business aims;
- **buying,** which involves understanding requirements, investigating options, making enquiries and placing orders;
- **expediting,** to ensure timely delivery of the materials etc. ordered, obtaining information on the status and progress of orders, and gathering intelligence about suppliers and the market in general;
- **inspection,** ensuring that suppliers conform to the specifications incorporated in orders placed;
- **administration** of these activities to ensure that they are carried out cost-effectively in the interests of the company's overall business aims.

The organization of the purchasing department in a large company follows the normal hierarchical structure. Usually the purchasing department manager has the same status as other department heads, and quite often will be a board member.

When a project task force is formed, the purchasing department will assign individuals to the task force in much the same way that the discipline engineers are assigned from their specialized departments. In this case, the purchasing specialists are responsible to the project manager for their work *on the project*. But they remain responsible to the purchasing department manager for the general professional standard of their work.

BUYING

In large organizations – whether corporate or task force – a basic decision is whether to centralize buying or to concentrate on localized buying, or to plan to use both. The advantages of *centralized* buying are as follows:

- a single large buying function requires fewer people than several small units dealing with the same volume of purchases;
- purchasing policy is more easily standardized;
- the centralized function will be able to place larger orders, with better opportunities for negotiating preferential terms and probably better service from suppliers;
- a common data bank on suppliers can be set up.

The advantages of *localized* buying are as follows:

- transportation costs are reduced;
- more prompt response to emergencies;
- closer personal relationships, with more appreciation of problems;
- earlier warning of impending changes;
- cost considerations for 'once-off' orders;

In practice, the choice of buying arrangements will be guided by considerations such as:

- the size of the order;
- the duration of the supplier relationship envisaged;
- the qualitative nature of the material or service to be ordered (e.g. whether the purchase order may involve technical advice);
- similarity between different orders;
- distance between the buying organization and the suppliers.

In large-scale industries there is a tendency to prefer centralization, at any rate for the most important purchases, in order to build up long-term relationships between the major suppliers and the major purchaser organizations. This contributes to stability of supply and consistent quality. It also promotes savings through 'just-in-time' (JIT) materials management methods, by reducing the need to maintain large inventories in storage.

MATERIALS MANAGEMENT

In manufacturing companies the JIT concept (i.e. organizing deliveries so that raw materials, components and other resources arrive 'just in time' for manufacturing to proceed without interruption) may be implemented by organizing the buying function so that it is integrated with manufacturing operations. This is logical when:

- the cost of materials is a large proportion of turnover;
- buying in ready-made components is preferable to making them;
- inventory control, cost control, purchasing documentation, computer-aided design and manufacture and related activities can be integrated in a single management information system.

Organizing the buying function in this way imposes changes in the company's relations with contractors and other suppliers, as well as in interdepartmental relationships. Such changes are worth making if the conditions mentioned above are satisfied by the nature of the company's business. But the integrated materials management structure seems to be less well suited to projects because of their 'once-off', time-limited characteristics.

ADMINISTRATION OF BUYING

The purchasing process begins with the issue of a *requisition*. We may consider a requisition to be a document instructing a buyer what to buy. It may be issued by anyone who has authority to commit project funds to the acquisition of project resources (e.g. the project manager or someone delegated by him). It is important that the requisition be clear and complete. The requisition should state:

- the name of the project;
- the name and position in the project of the person issuing the requisition;
- the name and account reference of the issuing organization;
- the name and classification code of the required item;

- the names and locations of other people/parties involved in the transaction;
- characteristics of the item, e.g. overall size(s), weight(s), etc.;
- a brief description of the item, its function and how it fits into the project;
- the item's location in the project schedule, and the required date and place for delivery;
- technical specifications and descriptions (for non-standard items), or standard reference(s) (for standard items);
- name of any statutory body involved in certifying the item;
- list of applicable national, company or vendor standards;
- drawings, diagrams, photographs and samples which help to define the item precisely;
- identification of potential suppliers.

These data allow the buyer to put together a formal *enquiry* to be sent to potential suppliers. The buyer has several options to consider:

- If he has purchased such items before, he may decide that there is no point in going out for competitive tenders, intending to purchase from the previous supplier. In this case, the buyer must get agreement from the requisitioning authority (and should record that agreement!).
- If he has a reliable, up-to-date data bank, he may choose a short-list of potential suppliers from the data bank.
- Alternatively, he may use trade directories as a basis for finding potential suppliers.
- He must decide how much of the requisition to include in the enquiry document, taking account of any considerations of commercial secrecy relating to the project, but taking care not to omit essential technical information.
- He must provide suppliers with (1) general conditions of doing business, (2) conditions specific to this purchase, e.g. inspection and test requirements, arbitration of disputes, etc.
- He should propose the method of payment for the purchase.

When the purchase is to be based on competitive tendering, enquiry documents should be designed to elicit strictly comparable bids. All the tenderers should receive the same information. We may, however, receive one of three possible replies:

- the bidder *declines* to bid
- the bidder makes a *direct* bid
- the bidder sends a *non-compliant* bid.

If the bidder declines to bid, the fact should be recorded and, if we can find out, the reasons should also be recorded for future reference. Direct bids are usually evaluated by setting out a list of criteria and assigning weighted scores to each bid against each criterion in much the same way as evaluating contractors (see Chapter 12). For major or critical items, technical specialists should be involved in the evaluation process. Non-compliant bids are reviewed with the requisitioning authority (and, possibly, technical specialists) to try to determine the reasons underlying the non-compliance. If the bid indicates some previously unrecognized advantage, it should be considered. Otherwise, non-compliant bids are rejected.

The *tender* is the first legally binding document in the purchasing process. If it is revised, the latest revised issue cancels all previous issues, and the latest revision becomes the binding document. The tender typically commits the supplier organization to supply specified goods and/or services at a specified time. The preferred tender is accepted *via* an order. The *order* typically commits the purchasing organization to pay a specified price at a specified time or event.

Regrettably, few purchases are actually made under such straightforward arrangements. Study any 'standard' conditions of sale or purchase from any large organization to see why!

An order is written evidence of the existence of a contract between the supplier and the buyer (representing the purchasing company). The buyer's order should contain everything that is contained in the seller's offer, but nothing more. (If it demands something more than was offered, it is merely an enquiry in bad taste.) The order must therefore contain the complete definition of the supply, including inspection arrangements, instructions regarding documentation to accompany the supply, the time and address for delivery, terms and methods of payment and part payment, the penalty for failure to comply with the contract, and the charge rates for any changes initiated by the buyer. A repeat order for a stock item may be standardized and straightforward. On the other hand, an order for a complex change may run to several volumes of text and drawings. Nonetheless, the principles embodied in the order remain the same.

Order documentation takes many different forms, depending on the nature of the purchase and the corporate culture of the purchasing organization. However, all orders specify:

- the name and address of the purchaser;
- the name and address of the supplier/vendor;
- reference to the enquiry and the final tender;
- date of the order;
- address for delivery and delivery date;
- specification and description of the item ordered (e.g. the same documentation as the enquiry);
- instructions about packaging, transportation and the documents which should accompany the item on delivery;
- instructions for communications within the contract;
- the price, with instructions for invoicing;
- the signature of the person authorizing the purchase.

EXPEDITING

The basic function of expediting is progress chasing. Expediters visit suppliers' offices, factories and workshops checking on the progress of work in hand so as to ensure that deliveries are made on time, or, if not, that early warning of delays can be built into project schedules. Expediters consequently become very important sources of information for buyers.

They are also highly valuable to project engineers because of their ability to mitigate problems caused by design (or other) change. This follows from the expediter's capacity to act as an informal link between project engineers and vendors. They are often able to catalyse discussion of possible technical options in such a way as to reduce the necessity for contractual adjustments, while still achieving the optimal solutions to technical problems that engineers continually aim for. And they may also be able to promote improvements (mutually by purchasers and suppliers) in quality management procedures so as to reduce the tendency of inspection to cause project delays.

INSPECTION

Inspectors check the quality of materials and workmanship at suppliers' work sites. They work to detailed quality specifications provided by the purchaser's engineers, but they are usually responsible for their day-to-day work to buyers and expediters.

The main areas of purchasing activity where inspection decisions have to be made are as follows:

- material certificates are to be delivered with the supplied goods: no inspection of the goods to be carried out;

- material certificates are to be inspected before the goods leave the supplier's premises: no inspection of the goods to be carried out;
- visual inspection of the goods – this comprises complete test and measurement of all items; selection of samples for inspection; and inspection of a predetermined sample of critical elements (e.g. of welds in a pressure vessel);
- on-line inspection during manufacture – this is the most stringent form of inspection, and is based on methods selected for the individual items to be inspected.

Most inspection methods are governed by national standards or company codes of practice which are well understood by the major suppliers. Critical items may be inspected under the supervision of international insurance agencies.

Inspectors are often undervalued by their employers. Yet employing well qualified and competent inspectors can have a considerable influence on project schedules by reducing, or eliminating altogether, delays arising in the suppliers' workplace. It is important, however, that inspectors are given authority compatible with their competence, so that they can take appropriate decisions themselves.

PURCHASING RECORDS

A good purchasing records system is a valuable, if not vital, means of improving the efficiency of the purchasing function. A suitable system should contain:

- **Current order files,** listing requisitions; competitive bids; purchase orders; progress records (delivery dates, milestones, expediters' reports, inspections, delays and delay recoveries, knock-on effects, concessions); change orders; final inspection reports; and closing statements.
- **Closed order files** – these provide the basis for comparing vendors for future purchases, using, for example, tender evaluation data and other intelligence information.
- **General ledger** – total value of purchases analysed by, for example, time period; project and project value; cost of processing each purchase order; percentage difference between price paid and min/max offers; percentage difference between min/max offers; number of offers evaluated per enquiry; delivery times and milestones for various types of purchase; number of purchasing activities per time period.

Suitable records systems can often be obtained in the form of proprietary computer software. However, it is not difficult to develop one's own software. This will probably suit one's own needs more precisely, and be cheaper too.

PURCHASING COMMODITY ITEMS AND CAPITAL EQUIPMENT

Purchasing commodity items

'Commodity' here refers to standard, 'off-the-shelf' items or materials the specification of which does not change, and which are purchased frequently. We have two main options:

1. To purchase forward for our foreseeable requirements. Consider:
   (a) price advantage if prices are tending to rise;
   (b) cost of storage;
   (c) cost saving on transportation;
   (d) cost of tying up capital in the inventory which we have to store.

2. To purchase in order to meet current requirements. Consider:
   (a) price advantage if prices are tending to fall, or disadvantage if prices are rising;
   (b) flexibility if design changes are likely;
   (c) frees capital for operational use;
   (d) cost of placing many small orders;
   (e) vulnerability if supplies become unavailable.

Purchasing capital equipment

Here the most important considerations are:

- Will it work satisfactorily?
- Is it really the most suitable equipment for the job?
- Will it earn an attractive return on the investment?

Again, we have two options in addition to the option of buying new equipment. Consider:

1. Buying second-hand equipment:
   (a) may be much cheaper than new, but must be thoroughly tested;
   (b) life expectation must be assessed independently of vendor's claims;
   (c) value of vendor's guarantees to be evaluated.
2. Leasing equipment:
   (a) advantage of smaller immediate payments, but may be more expensive in the long run because of high interest charges;
   (b) easy to obtain most up-to-date equipment;
   (c) owner may be willing to maintain it;
   (d) implications of owner keeping control need to be assessed.

CHANGE ORDERS

A *change order* may be defined as 'an order which changes any part of another order in force'. A change order has the effect of cancelling the order in force and replacing it with new instructions, although some parts of the original order may remain.

Change orders are almost inevitable in projects. In a development project, practically everything changes most of the time. In perhaps the majority of projects, people constantly think of changes which seem to be desirable. Sometimes changes arise because suppliers or owners relocate, or – for example, in JIT supply contracts – because the costs of deliveries change. In any case, arrangements for dealing with change orders should be made at the outset of the project.

Change orders arise because someone sees a need to revise something in the project as planned and authorized. These perceived needs are various and may disguise a hidden agenda. Consider the following:

- **The technology is improved during the project.** Changes are intended, for example, to reduce eventual operating costs sufficiently to pay for any consequent delay and additional expense. But this reason may disguise the need to get a change order for something quite different, such as the need to rectify design/construction mistakes (see following point).
- **Rectifying design or construction mistakes.** Contractors and consultants seldom discuss this, but most project engineers know mistakes are common enough. Sometimes a supplier can correct his mistakes himself. If, however, they affect other parts of the project, a change

order will be required. From the owner's point of view, it is important to ensure (1) that the change will enable the planned project objectives to be achieved, and (2) that responsibility for the cost of the proposed change is fairly allocated: no-one wants to pay for someone else's mistake.

- **The business environment of the project changes.** Market demand for the output of the project may increase or decrease significantly during the project. Revising design in the early stages of the project is usually neither difficult nor expensive. Late changes are both painful and costly to the owner, while suppliers expect to be paid for the work they have already done. A somewhat similar situation occurs when the regulatory environment of the project changes (see following point).
- **The regulatory environment of the project changes,** e.g. under the impact of new regulations governing worker health, safety or environmental conditions. Typical examples relate to reducing emissions; including noise emissions, and changing the layout of vessels and tankage.
- **Budget shortfall.** If project funds run out, part of the planned project work must be cancelled or postponed. Cancellation or rescheduling requires change orders. There will be ample scope for disagreement between owners and suppliers over the value of work completed (and due to be paid for) and the reasonable (or otherwise) expectations of suppliers for compensation. Postponement of deliveries almost always affects prices and may affect premiums for critical delivery items.

These observations make it clear that the market can provide opportunities for suppliers to benefit from change orders, especially if the owner is not master of the technical details of his project. For example, because the lowest price attracts the most attention in competitive tendering, suppliers are motivated to offer the minimum that meets the owner's specification. An unscrupulous supplier may cut his margin, taking the chance of making profit on overpriced change orders when the owner is trapped. It is a winning chance when the judgement of accountants has greater weight than the judgement of engineers in evaluating proposed changes.

Much turns on the validity of the owner's specifications. Scrupulous suppliers accept that they are responsible for their supplies being suitable for the purpose for which the supplies are sold. If they consider that the owner's specification is deficient, a scrupulous supplier will explain his reasons for not complying with that specification (because if a compliant supply does go wrong, the supplier must make good at his own expense).

A change order comes into effect following its approval by the project manager or someone else to whom approval authority has been delegated. It gets approval because its benefits, however assessed, are judged to outweigh its costs. But what is the cost of a change order? This is the sum of:

- the cost of the work done which has to be replaced;
- the cost of design or other work done in order to recognize the need for the change order;
- the cost of preparing for negotiations;
- the cost of negotiating with suppliers;
- the supplier's extra charges;
- the cost of the 'knock-on' effect of the change, when implemented.

The knock-on effect includes disputes about the necessity for, or the responsibility for the cost of, changes. Disputes arise because change order and change control procedures are not in place, are inadequate, or are misused. In the worst case, disputes lead to litigation, from which no-one emerges satisfied except the lawyers.

A general procedure for considering change is as follows:

- identify the items to be changed;
- identify the items which will be affected by the change;
- work out the cost (see above) for possible alternative approaches;
- circulate the information and invite comments;
- make the choice on technical grounds using clear, unambiguous criteria;
- give the purchasing manager the opportunity to make the best deal he can for the chosen option.

Change orders inevitably result in loss of time, and usually loss of money from the project budget. We need to know where this goes: *cui bono?* Who gains from change orders? If we can see the cost of a change order as a good investment, then the change is probably well under control.

14.4 Summary

The consequences of quality failure can be very serious. It is unacceptable to allow situations to arise in which substandard quality becomes apparent only when something fails in service. Consequently, each and every activity which directly affects 'fitness for purpose' must be systematically monitored and controlled to ensure its own fitness for purpose. QA and QM provide both the assurance and the demonstration that this has been achieved. This is neither a separate discipline nor an add-on. It is a planned, methodical, procedural approach to work, which is an essential tool of project management.

The efficiency of purchasing depends very much on teamwork, and in particular on the quality of the instructions provided by project engineers to the buyers, who initiate purchasing activities. It is essential to define clearly the responsibilities and the authority of the buyer.

The purchasing department and the purchasing team members will be most effective when they can develop long-standing relations with their suppliers – relations which should be based on the objective assessment of suppliers' capabilities and straightforward dealing.

In the circumstances of the competitive market, business aims and business organizations change. The purchasing function is in the forefront of these changes and can make a powerful contribution to meeting the challenge of business change provided it has the committed support of the project leaders.

15
Managing Project Control

15.1 Introduction

Controlling the project so as to complete it to specification, on time and within budget is at the heart of the project manager's job. Many people – especially authors of books on project management – believe it is also the rest of the body and the spirit of the job too, so it is very easy to find detailed descriptions of the techniques used for project control. If you are interested in the details, you can find them in any project management manual. Detail apart, however, it is important for project leaders, as well as project managers, to appreciate (1) the existence of these techniques, and (2) the ways in which they can – and cannot – be usefully applied. But, perhaps most importantly, to understand how they can help to support effective project policy and decision making.

15.2 Work breakdown structures

BREAKING DOWN THE PROJECT

A work breakdown structure (WBS) consists of all work and all costs collected in a hierarchy with no omissions and no duplicates (Fig. 15.1). It does not much matter how the hierarchy is arranged. A WBS may break the project down by groups of items, by types of work, or by work sites, provided everything is included and made visible.

An example of a WBS for an oil refinery, broken down by process functions, would be:

- system – oil refinery;
- process group – crude distillation unit (there will be several other process groups in the system);
- process unit – crude column (there will be several other process units in the group);
- equipment item – column shell (there will be other equipment items in the unit);
- related bulk item – column piping (there will be other bulk items related to the equipment item);
- and so on.

Each item in the WBS must have work done on it to turn the raw inputs – or project 'receivables' – into the operational facilities required – the project 'deliverables'. For example, each item must be designed, purchased, constructed and commissioned. These work functions can also be broken

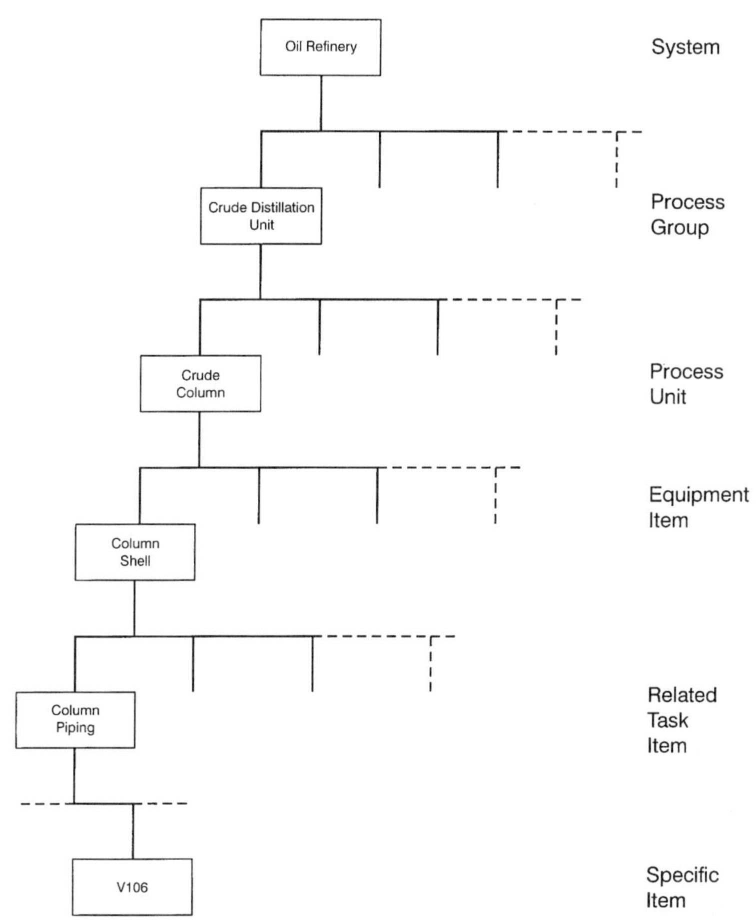

Fig. 15.1 The work breakdown structure. This makes important tasks visible; allocates responsibility to a named person (who owns the 'task'); and is the basis for cost estimating and cost control. It is not very important *how* the WBS is broken down, provided it includes all work and all costs, with nothing omitted and no duplicates.

down to increasingly detailed levels: process engineering, discipline engineering, drafting, purchasing, inspection, and so on.

The manhour requirement for each hardware item, and the manhour rates for each function, plus the cost of any special resources such as computer time or special construction equipment, when added to the cost of the hardware item itself, produce the cost estimate for the deliverable (see Chapter 8).

The breakdowns of hardware and work operations (or 'functions') can conveniently be combined as a matrix (Fig. 15.2), which can be developed for any level of WBS to show all relevant project activities, which can then be:

- matched to the project organization
- delegated to named responsible individuals
- aggregated into commitment packages.

The level of detail in the WBS determines the number of project activities we identify. In the early stages of project development, when we are dealing with concepts, the number will be

Project: NNN	Functions						
Hardware	Design	Purchase	Construct	Commission	Other	Proj Mgt	Total
Item 01							
Item 02			$xJoe				
Item 03							
Item 04							
Total							

Fig. 15.2 In the hardware/function matrix, the budget for constructing item 02 is $*x*, and Joe is the 'owner' of the task to construct item 02 within budget and schedule (and according to specification).

small. As the project becomes more completely defined, the number of project activities will grow, rising to orders of magnitude of 10^4 to 10^5 for a large project. But for direction and overall management, the details will be aggregated again, and we will deal with relatively small numbers of project activities.

It is the project manager's task to decide whether a project activity has been defined sufficiently in the WBS. The decision has to be based on his judgement of the circumstances, his experience, and his 'feel' for the demands of control for *this* project. Typically, WBSs for project control are prepared at three levels of breakdown:

- **Level 0** – management overview. Deliverables are expressed in terms of projects (in a multi-project environment) or process groups.
- **Level 1** – project manager information. Deliverables are expressed in process groups and process units.
- **Level 2** – project engineer information. Deliverables are expressed in process units and item deliverables.

Reporting should allow each level to be fed upwards to the next higher level.

SINGLE POINT RESPONSIBILITY

Someone must be responsible for each project deliverable. Underpinning WBS is the concept of single-point responsibility (SPR). At any time, *one* person is nominated to be responsible for the realization, on time and within budget, of a specified project deliverable. When the specified item is the project as a whole, this person is the project manager. The breakdown of project responsibilities must match the WBS, so that the WBS plus SPRs form the basis for controlling all project activities.

15.3 Cost, time and resource control

DOCUMENTATION

For each project activity identified in the WBS, we can prepare a record showing:

- the *name* of the activity and a code representing its unique identification;
- the *scope of work* required to complete the activity;
- the *resources* required to complete the scoped work;

- the *cost* committed;
- the *time* required;
- the *individual responsible* for ensuring satisfactory completion of the activity.

In addition, the record should show the sources of all estimates relating to the activity, the assumptions made for the respective estimates, and a cross-reference to the appropriate budget item or expenditure authorization. Records such as this, sometimes called cost-time-resource (CTR) sheets, can be collated into a file (called the CTR catalogue) which becomes the primary management data base for controlling the project.

If the CTR data base is to be used effectively, the coding of activities must be carefully designed, because cost and commitment control and cost reporting are all based on the activity code. The activity code also allows invoices to be automatically identified in the financial reporting system. The activity codes must be designed to be consistent with existing company procedures. Otherwise the cost of re-coding when the project is eventually handed over to the business or operations function can be astronomic.

COST AND COMMITMENT CONTROL

When the project begins, it starts to incur costs: manpower costs, purchases of equipment and materials, deliveries and storage. Requisitions, tenders, orders and invoices fly about the project office like demented hornets, and with as much potential for causing confusion and pain. Very many transactions take place in a very short time.

It is important to distinguish *committed* from *actual* costs. When something is ordered for the project, we are committing its cost, although the actual cost may not appear in the accounts until some time later, when the invoice is paid. From the project point of view, it is vital to record *commitments* in relation to the project schedule.

When project accounts are made up, a potential source of dismay is due to accountants, whose sole ambition is to get an accurate picture of costs *now* (i.e. actuals) because they see their life's work in terms of stewardship. Project managers want to know their costs *now* and in the *future* (i.e. commitments) because they see their function in terms of control. A special breed of accountant has evolved in the major project environment to resolve this 'cultural' problem. Project accountants are more in tune with the special needs of administering funds in the interests of the project.

The first step in controlling project commitments is to make a commitment plan. This should be part of the project execution plan, which would have been sanctioned when the project was approved for implementation. The budget for this is based on a *control estimate*, and the total amount of the commitment plan must be within the approved budget.

It is useful to prepare a preliminary commitment plan early in the definition phase. It provides a basis for developing the final commitment plan, and can be used to support requests for the approval of commitments which must be made in advance of the final commitment plan, e.g. when contractors or consultants are retained to develop the project specification.

The WBS provides the means of effecting control within the commitment plan. Commitments comprise one or more project activities, documented in terms of project deliverables, which are the fundamental units for managing project planning, cost estimating, cost control and contracts.

Commitments are normally authorized through a formal request procedure to ensure that only commitments which conform to the commitment plan are approved. The documentation of commitment request and approval can provide a very important control mechanism, based on the following principles:

- **before commitment:**
  - the project must be covered by an approved budget;
  - the project implementation plan must have been approved;
  - commitments must be expressed in terms of deliverables in a WBS as described above;
  - the value of each commitment must not exceed the value of the respective request submitted for approval;
  - the total of approved commitment requests must be less than the total of the approved budget, remembering that only corporate management can release the overrun allowance.
- **after commitment:**
  - the value of the commitment, of the work done, and of the expenditure must be recorded systematically for control purposes;
  - the value of the work done must be less than (or, on completion, equal to) the value of the commitment.

In practice, applying these principles to project control is made much easier by adopting a consistent system of coding for hardware items and work functions throughout the WBS. The coding should be the same as that used for financial accounting, so that commitments, activities and invoices can be reconciled. The coding should also be compatible with existing operational coding systems, to avoid the problem of re-coding when the project is handed over to the owner organization.

15.4 Planning and scheduling

On taking up his assignment, a project manager makes a commitment to the future. Because time travel is strictly one-way (except in science fiction) he worries: (1) has he done everything today to meet the future commitment?; (2) is there enough time left to deliver his promise?; (3) while he knows what he has done thus far, he can't look into the future; so (4) his decisions have to be based on history and guesses. To mitigate his worries, he has *planning* and *scheduling*. These are different, but related, concepts of the project control task. Planning sets out what has to be done in order to achieve the project's objectives. Scheduling adds to the plan the time and resources required, from which suitable control mechanisms can be developed.

Planning begins with a definition of the project objectives, ensuring that all the parties engaged in the project understand them. Next, corporate and project strategies need to be determined, and the project philosophy worked out, taking account of constraints and limitations. The major project activities are identified using a WBS (level 1), and the project logic governing the interdependency of project activities is established.

Scheduling begins with information for project activities, starting at the level of detail consistent with the WBS for the major activities, and working up to the management level. (It is dangerous to try to schedule from a less to a more detailed level.) Data for each activity are recorded on the respective CTR sheet, and the sheets are then collected together to form the CTR catalogue.

Scheduling reflects the WBS at the respective project phase. It shows the key dates and durations of all the identified project activities. Scheduling provides the understanding, the means of getting approvals and the means of monitoring and controlling the project. It also helps to pinpoint and assess project risks. The risk of unforeseen delay is one of the main risks leading to project failure.

NETWORKS

We can immediately see that some activities logically depend on others. Some activities can begin only when others have finished. Some must start or finish together. Some may be completed only after others have started. These relationships can be represented by a network which shows the project activities, their individual durations and dependencies. (A 'dependency' is simply the relationship between one project activity and another with which it is linked for project purposes.)

The construction, interpretation and use of networks form a main topic in most project management text books. There are also many software systems for building and analysing networks which include self-teaching modules, so that anyone who wishes to learn has no shortage of sources. For this reason, only an outline of network techniques is given in this book. If you are completely unfamiliar with the concept, an elementary discussion will be found in Appendix 3.

Project work activities and relationships

The most important interrelationships between project activities are, for example:

- Activity A (the 'predecessor' activity) must be completed before activity B (the 'successor' activity) can begin. This is the most frequently used project relationship. This is *'finish-to-start' (FS)*.
- Activity B cannot begin until activity A has started, although the respective completion dates may be different. This is *'start-to-start' (SS)*.
- Activity B cannot be completed until activity A has finished, although the start dates may be different. This is *'finish-to-finish' (FF)*.

Various conventions for drawing networks exist – see any textbook on critical path analysis and project evaluation and review technique (PERT) – and we need not discuss them here. The purpose of drawing the project network is, however, to determine the *critical path* (or paths: there may be more than one), i.e. the longest path representing the shortest time for completing the project. When the critical path has been found, you will find other paths through the network which are non-critical. These paths have time available for work which does not add to the shortest completion time. This is *(positive) float* time, which can be used to reduce time pressure on the use of resources elsewhere in the project.

Usually, work on a successor activity is scheduled to start immediately after the predecessor activity has been completed. This is the 'as soon as possible' (ASAP) mode of scheduling. An alternative is the 'as late as possible' (ALAP) mode, when the successor is scheduled for completion as late as permitted by its free float time (i.e. the amount of time that a non-critical successor activity can be delayed without affecting the start of another activity).

ASAP scheduling provides flexibility by keeping free float available. This is the reason for using the ASAP mode wherever possible. On the other hand, if we scheduled all the activities in a project using ALAP, they would all lie on the critical path, so ALAP should be used only when an activity *must* be started as late as possible.

Scheduling relationships may be adjusted by using *lag* or *lead* times. A lag time is a delay introduced between the completion of a predecessor activity and the start of its successor. Lead time (or 'negative lag') is an overlap between the completion of a predecessor and the start of its successor. Introducing lead time can sometimes help to shorten the overall schedule of the project by increasing overlaps (e.g. in 'fast-tracking'), but must be used with caution.

Usually, the project will have a given start date, and the schedule is calculated forward as work is defined and dependencies worked out, until the overall completion date is arrived at. This is *forward scheduling*. Sometimes, however, the project completion date is the specified 'given'. In this case the schedule must be calculated backwards in time to establish the required start date.

This is *reverse scheduling*. When the required schedule has been established by reverse scheduling, with the project completion date as a 'milestone' at the end of the project, it is good practice to revert to forward scheduling for project control, because this will indicate negative float if, for example, changes to the plan would cause the required completion date to be missed.

Project schedules may be constrained by activities being *date-dependent*, or by the completion of an activity being *required by a specified date*. Date-dependence specifies the earliest date that an activity can start. The date required specifies the target date for completion of the activity. If the schedule forces a completion date later than this target, the amount of time by which the target is missed is known as *negative float*. Negative float indicates when the activity should be started in order to meet the target date required.

Management and networks

At the management level, simplified networks may be used to get a 'feel' for the project. Guidelines for using simplified networks are as follows:

- keep the number of activities as small as practicable, by aggregating activities where this is feasible;
- keep the schedule simple, but take care not to lose significant linkages between project activities;
- check the resources needed (manpower, equipment, etc.);
- analyse the network to find the critical activities (i.e. those lying on the critical path) and non-critical activities where 'float' is available;
- revise and optimize. If the network does not result in the required completion date, it must be recast. For example, sequential activities may be rearranged to run in parallel wherever this is practicable, or 'float' time can be reallocated. Of course, you may find that there is no way to achieve the required completion date with the resources available. In that case, you must either get additional resources, or accept that the project schedule will be longer than expected.

Detailed network analysis can be very time-consuming and tedious, so much so that for all but very small networks it is done using special computer software. However, useful information can be obtained from networks simplified so that only about 10 major activities are considered. Such a network can be drawn and analysed manually, although if you want to ask 'what if' questions to test several alternative options, manual analysis, even of very small networks, becomes excessively tedious. For this, or for more activities, a PC is essential. For level 1 purposes, about 150 activities is a practical maximum.

All the data we use in a network – i.e. the durations of activities – are estimates. They are not certain, so that they should preferably be treated as probability distributions in a similar way to the treatment of costs and revenues in the economic appraisal of projects (Chapter 9).

Interpretation of deterministic networks tends to be optimistic. Suppose two activities, A and B, must be completed together before the next activity, C, can begin, and both A and B could take 2 or 3 time units to complete. Then one of the following applies:

- if $A = 2$ and $B = 2$ then C can start after 2 units;
- if $A = 2$ and $B = 3$ then C can start after 3 units;
- if $A = 3$ and $B = 2$ then C can start after 3 units;
- if $A = 3$ and $B = 3$ then C can start after 3 units.

The average for A and B is 2.5 units; the average for C is 2.75 units. People tend to take the average for A and B as signalling the start time for C, instead of the average for C. This is 'nodal bias'. In the example, the nodal bias is:

$$\frac{(2.75 - 2.5) \times 100}{2.5}$$

i.e. 10 per cent.

The larger the number of activities, the greater the nodal bias (i.e. the tendency to unrealistic optimism). Many subcontractor activities in the network will tend to increase nodal bias. Fast-tracking increases the number of parallel activities, and therefore increases the risk of flawed time estimates due to nodal bias.

Probabilistic analysis of networks requires computer software. You can, however, get a 'feel' for the impact of probabilistic estimates of durations by substituting crude (weighted) averages for the deterministic durations. Consider, for example, Table 15.1. In the early stages of planning of some projects (especially R&D projects), not only the activity durations, but also the dependencies are probabilistic, making analysis extremely complex.

Table 15.1 Estimating probabilistic durations

Deterministic duration	Minimum/maximum duration	Minimum/most likely/maximum duration
10	9/13	9/11/15
	(9+13)/2	[9+(4×11)+15]/6
Probabilistic 'average'	11	11.3

Bar charts, 'milestones' and progress measurement

Bar charts (also called Gantt charts after their originator) show the durations of project activities and can be adapted to indicate activity dependencies (*see* Fig. 15.3). 'Milestones' mark the accomplishment of key project activities and are often used to trigger payments to contractors. In a network they can be specified as activities of zero duration. They should be specified carefully, referred to the completion of defined activities and not to calendar dates, because in the latter case they need not bear any relation to actual work done!

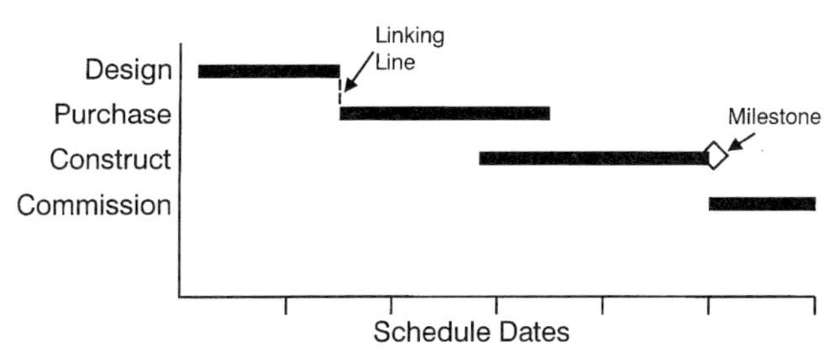

Fig. 15.3 Bar charts show the duration of the project tasks. Linking lines can be used to show dependency between tasks. Milestones mark the accomplishment of key tasks.

Milestones can be defined by grouping CTR sheets. If contractors' plans and CTR groups are incorporated into the contract, the milestones so defined become a useful instrument for controlling contractors' schedules.

For control purposes, progress should be measured in '100 per cent complete' steps. This requires the WBS to be broken down to a level where deliverables can be referred to '100 per cent completeness' (i.e. there will be many, small progress steps – see 'Earned value analysis' below).

'S' curves

These curves relate project cost (commitment or expenditure) or percentage completion to project time. S curves are quite characteristic of the evolution of project development: a set of S curves is perhaps the most fundamental representation of the project plan (Fig. 15.4).

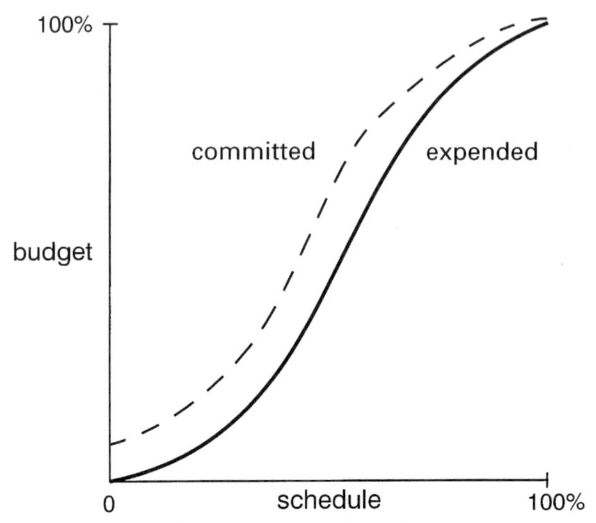

Fig. 15.4 'S'-shaped curves are characteristic of the planned relationship between project budget and project schedule.

Manhours worked

This records the actual work done on the project. It is compared with the manhours planned to have been accomplished at the date of the record. Accurate measurement is often difficult, but it is far more useful to have even approximate information promptly (so that timely corrective action can be taken) than to have very precise information too late for effective action.

Resource levelling

The resources actually required for a project, either in manhours or money, are often found to exceed the resources available at the time when they are needed to complete specified project activities. At other times, more resources may be available than are actually needed. Identifying 'float' time is a pointer to this situation. In this case, we have to renegotiate or reallocate the resources available and/or the timing of their availability. This process is called 'resource levelling' and is a vital aspect of project management (*see* Fig. 15.5). It is not easy to do manually, even when only a few project activities are involved. Fortunately, computer software is readily available to help. But remember the axiom: GIGO (garbage in means garbage out!).

Constraint handling

This is a relatively new software-based technique for generating optimal solutions to problems that are frequent in project planning and scheduling, such as allocating scarce resources to several project activities where some activities are required to share the same resource, but the sequence of activities is not specifically ordered. The problem thus compounds the cumulative resource constraint with a sequencing constraint, a situation which cannot be handled efficiently by conventional project management methods.

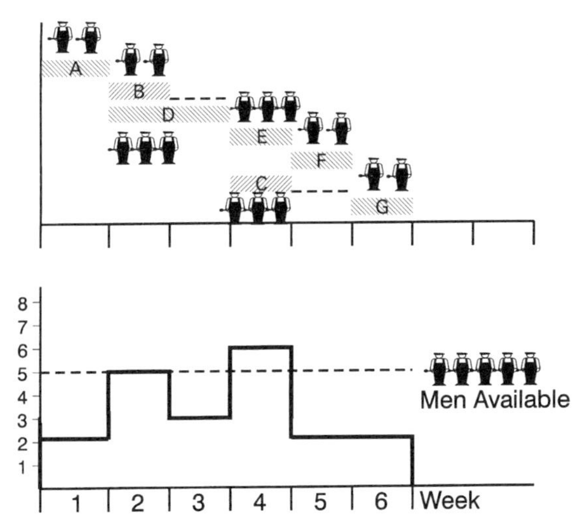

Fig. 15.5 Resource levelling. The schedule requires too many men in week 4. We need to rearrange the tasks in order to 'level' the resources.

EARNED VALUE ANALYSIS

Also known as 'performance analysis', this is a method of relating the cost and time of work actually accomplished on the project to the cost and time planned. Three special items of information are needed:

- a measure of the work actually done, including indirect costs (e.g. the cost of project management);
- the actual cost of the work done expressed in cost units per work unit;
- the budgeted (planned) cost of the work done, expressed in the same units.

All this information must be referred to the same moment in project time.

The relationship between work done and time is always expressed in terms of the budget. This allows different kinds of work to be aggregated. A key underlying idea is that 100 per cent of the budget is the *maximum* that can be earned (although, of course, the budgeted *cost* can be exceeded!). Using these measures, we calculate:

- the actual cost of the work performed (ACWP);
- the budgeted cost of the work performed (BCWP);
- the budgeted cost of the work scheduled (BCWS).

We then plot these on the current time axis (or the time when the measures were made) of the project S curve (Fig. 15.6). It is usually easier to get reasonably accurate measures of work performed when the project activities are specified to be of short duration. Then:

- BCWP – BCWS is the 'schedule variance'. A negative answer indicates that you are behind schedule (in money terms).
- BCWP – ACWP is the 'cost variance'. A negative answer indicates that you are overrunning the budget.

The plot should be updated regularly so as to develop the earned value trend over time. Then, considering these variances together with the current possibility of resource levelling allows us to see what can be done to bring the project back to plan or, if that is not practicable, what the

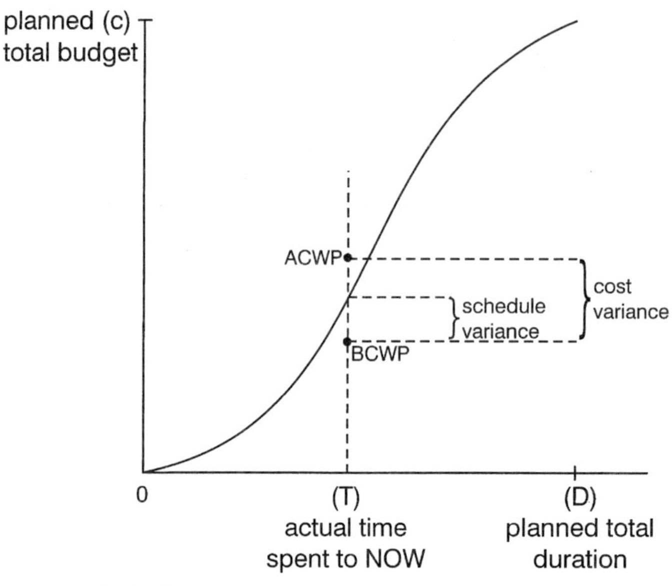

Fig. 15.6 Basic information for earned value analysis.

likely cost of overrun and/or delay to project completion may be (so that corporate management can adjust their strategy accordingly).

Two ratios are also useful for control purposes:

- schedule performance index (SPI), which is BCWP/BCWS;
- cost performance index (CPI), which is BCWP/ACWP.

Forecasts to completion

Let the planned total project budget for completion be C; the planned total project duration at completion be D; and the actual time spent to date on performing the work be T. Then the estimated time to completion is ACWP + (C – BCWP)/CPI, and the estimated cost to completion is T + (D – T × SPI))/SPI (Fig. 15.7).

Earned value analysis is so neat that it deserves to be very popular. It isn't, for the following reasons:

- It needs costs expressed as 'committed' rather than 'spent' costs. This is unfamiliar ground for many accountants, whose cost codes are often unsuitable for committed cost reporting.
- It is difficult to measure work actually performed, as distinct from identifying the money spent on the work.

Consequently, the project WBS should be broken down into components small enough so that each component can be assessed as either 0 per cent or 100 per cent complete. This is tedious, but avoids the situation where a contractor can claim 90 per cent payment against 90 per cent completion, while the remaining 10 per cent is the critical part of the job.

15.5 Change control

A change is any alteration to an agreed project plan. No matter what is done to 'freeze' plans, changes inevitably arise in the course of a project. Controlling them is probably the most important factor in ensuring project success.

Suppose, in Fig. 15.6 we have:
Planned total project budget, C = $1 000 000
Planned total project duration, D = 100 weeks
Time expended up to NOW, T = 50 weeks.

and, at time NOW, BCWS = $420 000
BCWP = $400 000 (we are behind schedule)
ACWP = $450 000 (we are over budget).

When can we expect to finish the project, and at what cost, if work progresses at the present rate?

$$SPI = \frac{400\,000}{420\,000} = 0.95$$

Estimated time to completion $= 50 + [100 - 50 \times 0.95]/0.95 = 105.3$ *weeks*

$$CPI = \frac{400\,000}{450\,000} = 0.89$$

Estimated cost to completion $= 450\,000 + [1\,000\,000 - 400\,000]/0.89 =$ *$1 124 200*

If we repeat the analysis after we have taken some corrective action, we shall be able to see the effect of the action taken in terms of revised estimates of time and cost to completion.

Fig. 15.7 Earned value analysis used for forecasting project outcomes.

When should a proposed change be considered? In the following cases:

- *if* it will make a significant improvement to the health/safety/ environmental (HSE) standards of the project;
- *if* it will significantly improve the project's technical performance;
- *if* it will significantly improve the chances of shortening the project schedule;
- *if* it will significantly improve the profitability of the project.

These improvements, when proposed, must be quantified and their real value objectively assessed.

When should a proposed change be rejected? In the following cases:

- *if* the project satisfies statutory requirements/standards without the change;
- *if* the project will meet its duty, performance and quality specifications within the budget and on schedule without the change.

There is plenty of scope for disagreement about particular change proposals. The weight to be attributed to the above six points in each case is largely a matter of corporate culture. Consequently, general policy for dealing with project changes should be established before the project is initiated. We can, however, set out some procedural guidelines here and now:

- change control procedures should be simple, and understandable by everyone in the project team;
- change control procedures should promote prompt, authoritative decisions;
- they should be flexible enough to accommodate the requirements and constraints of the various project phases (note that the later the change, the greater its effect on budget and schedule);
- any change proposal must be supported by fully documented analysis and appraisal of its likely impact on the project budget, schedule and performance;

- if a proposed change will affect contractual obligations, the consequences and implications must be agreed by the contracting parties before approval – variations to contract terms and conditions must be made only after approval has been given;
- approval for a change should never be given ahead of full documentation – as a minimum requirement, the change must be approved by the level of authority which approved the original plan;
- proposals for changes should be made in accordance with a formal procedure and endorsed by the line (operations) management who will be affected by the proposed change.

Change proposals are best made by raising a standard change control form. This is reviewed by a change control committee consisting of, for example, the project manager, the appropriate technical specialist, the project cost or planning engineer and a QA representative.

Because of the likelihood that a change will adversely affect the project budget or schedule or demand for resources, changes should be minimized where they cannot be avoided altogether. Rigorous change control procedures help achieve this. Good management practice also helps to:

- ensure that project development is based on the best available information;
- keep in close touch with regulatory authorities and interests which may be affected by the project;
- ensure enough time is available for preparing project specifications;
- make sure the project specifications and instructions are comprehensive and clear;
- ensure users are involved as members of the project team;
- make quality reviews an inherent part of the design process;
- carry out a critical review of major design decisions during design reviews;
- carry out regular reviews of the project.

After review, the project manager must check, before seeking approval, that he has budget cover for the change.

Another good reason for avoiding changes wherever possible is their deleterious effect on the project team's morale. However, it is rare for a project to go ahead without changes, so it is important to try to minimize their impact by careful attention to the details of project control.

15.6 Reports and approvals for project control

Reports

Reporting is a vital control mechanism. Project progress must be reported regularly and frequently enough that management decisions can be taken by the appropriate authority in time to be effective. Normally, this means reporting on project progress at least monthly. Reports should be unambiguous, indisputable, current and clear. Above all, they should result in decisions. Reports must state:

- whether project work is, or is not, going according to plan (in quantified terms);
- if the project is not going according to plan, why it is not;
- the remedies/corrective actions proposed, together with the consequences of the proposed action.

APPROVALS

Formal approvals are normally required:

- at each significant phase of project planning and development (the project does not proceed to the next phase until the current phase is approved);
- on receipt of statutory authorizations, where applicable (e.g. planning, HSE, certification of pressure or offshore equipment etc.);
- for each budget application (in relation to the several phases of project development);
- before entering into significant commitments (e.g. contracts or payments);
- at each significant stage of engineering design (e.g. as marked by issue of engineering documentation);
- at project handover;
- at project close-out.

Levels of authority for the various approvals, and approval forms (standardized wherever possible) should be established in corporate procedures before the project begins.

A basic policy for approvals is that:

- all information required by the approving authority must be provided, clearly and concisely;
- an approval for a particular activity or commitment can be requested once only;
- activities for which approval is requested must be clearly defined;
- SPR applies to obtaining the approval.

15.7 Summary

This chapter has briefly described the main approaches and methods used to control project cost, time and the utilization of resources. The methods are typical of those used for controlling large projects, but are applicable in principle to all projects.

Although the actual use of control methods falls to the project manager, appreciation of their use and their possibilities is also important for all the project leadership. These methods are, of course, not the only tools of project management which should be understood by project leaders: economic appraisal, contract strategy, project organization, information management and quality management are, among others, also important. But cost, time and resource control are among the most highly visible management tools and, if for no other reason, are obviously significant for those responsible for project policy making and direction. In fact, they embody the very basis of capability for project control:

- define project activities in a plan
- measure progress against the plan
- correct deviations.

You can't control what you can't measure. The problem in managing control is really choosing what to measure.

16

Environmental and Community Issues

16.1 Introduction

The coming into existence of a project disturbs the outside world – the environment and the community – in a variety of ways. Some of these impose costs which are difficult to measure and compare with the expected benefits of the project, yet which, in recent times, have been perceived as being very onerous – acid rain and global warming are examples.

With the increasing pace of worldwide industrial development, and the increasing number, scale and complexity of projects, the global community has taken these problems very seriously. Almost all countries now require projects to conform to legislation designed to reduce the project's impact on the environment and on the community (not simply of people, but of all 'living resources') affected by it.

The most general context for this legislation is that of 'sustainable development'. This idea was defined in the Brundtland Commission report of 1987 as 'meeting the needs of the present without compromising the ability of future generations to meet their own needs'. Putting it into practice has since been discussed in a series of essays (the 'Blueprint' series) the thrust of which is to show that achieving sustainable development will require a radical reorientation of industrial economies (Pearce, 1993).

Following the Brundtland report, the idea of sustainable development was adopted by the countries signatory to the UN Conference on Environment and Development (UNCED) held at Rio de Janeiro in 1992. The so-called Agenda 21 stated the signatories' intention 'to adopt a national strategy for sustainable development', the aim of which 'should be to ensure socially responsible economic development while protecting the resource base and the environment for the benefit of future generations.'

UNCED did not define what it considered to be socially responsible; what are the legitimate interests of future generations; how far *we* should protect the resource base; or what is meant by a 'resource base'. So there remains confusion between the technical characteristics of a particular development route and the morality or otherwise of choosing it (Beckerman, 1995). Thus, in the context of project development, it has been said (Little and Mirrlees, 1990):

> Sustainability has come to be used in recent years in connection with projects . . . It has no merit. Whether a project is sustainable (for ever – or just a long time?) has nothing to do with whether it is desirable. If unsustainability were really regarded as a reason for rejecting a project, there would be no mining, and no industry. The world would be a very primitive place.

Other issues are raised by the market economy in which a project exists. The free market is, by its nature, unstable. It is not well suited to dealing with the long-term problems posed by sustainable development. In a consumer-oriented market, demand is driven by want. In an economy oriented towards sustainable development, demand would be driven by need. In a market economy, technology is applied to maximizing throughput. For sustainable development, technology should be applied to maximizing efficiency. But whatever view we take, sustainable development is clearly a significant influence on project planning and implementation.

Broad guidelines on sustainable development have been published as a *Charter for Sustainable Development* by the World Industry conference on environmental management in 1991. The headings are quoted below:

1. Corporate priority
2. Integrated management
3. Process of improvement
4. Employee education
5. Prior assessment
6. Products and services
7. Customer advice
8. Facilities and operations
9. Research
10. Precautionary approach
11. Contractors and suppliers
12. Emergency preparedness
13. Transfer of technology
14. Contributing to the common effort
15. Openness to concerns
16. Compliance and reporting.

The aims of the charter are (or are supposed to be) brought into practice through national legislation. An organization which plans to develop a project in the course of implementing its business strategy must conform to whatever laws apply to the project location. Thus, in the UK, the Environmental Protection Act (1990) requires companies to minimize the environmental damage caused by their operations by adopting the 'best available technologies not entailing excessive cost' (known as BATNEEC). This approach is sometimes satirized as adopting the 'cheapest available technology not involving prosecution' (CATNIP).

An alternative approach, often claimed to be more cost-effective than BATNEEC operated with standards for permissible levels of pollution, is to apply pollution taxes. This, in effect, is to legislate so that the polluting organization pays a penalty commensurate with the damage caused by the pollution for which it is responsible. A refinement is the use of tradeable pollution permits, a concept which attempts to create a 'market' which would in theory distribute the costs of pollution abatement equitably. However, a major difficulty in implementing these ideas is: how should we assess environmental costs?

16.2 Environmental project economics

Our project development proceeds on the basis of series of studies intended to reduce uncertainty regarding the balance of benefits versus costs arising from the project's coming into existence and operation. Very typically, our eventual decision depends on our assessment of the present value of future cash flows (e.g. NPV or IRR, or similar cost/benefit indices – see Chapter 9). Quite apart from the impossibility of assigning a cash value to some features of the environment (such as an area of great natural beauty which happens to overlie an oil field), the effect of discounting over the very long periods of time implied by 'meeting the needs of the present without compromising the ability of future generations to meet their own needs' is to work against the best technical solutions to environmental problems.

Discounting at the rates typical for commercial project investment, when applied to environmental issues, assumes that we – the present generation – have exclusive rights to the world's environmental resources. This assumption runs directly counter to the aims of the

Charter for Sustainable Development. Thus, if we really intend to achieve sustainable development, it seems we have to use significantly lower discount rates or different appraisal methods when assessing the environmental aspects of out projects.

A consequence would be the need for a legislative framework such that projects would not be appraised on commercial market considerations or quantitative costs and benefits, but on some notion of ethics and equity in the distribution and consumption of environmental resources (Norgaard, 1992), perhaps in relation to an international agreement to set limits to those economic activities which cause most damage to the world's stock of environmental resources (Jacobs, 1991). But these seem to be ideals for the future, although they do now colour our perceptions of the place of projects in the environment (Moilanen and Martin, 1996).

16.3 Health, safety and the environment in day-to-day project management

As the project evolves, the project manager's responsibilities for health, safety and environmental (HSE) issues change. The following summarizes the main HSE issues that demand the project manager's attention during the phases of project development.

Fig. 16.1 BP exploration. The 85 000 tonne concrete gravity base for BP's Harding Project Field floating in the construction dock at Hunterston (April 1995). Photograph courtesy of British Petroleum.

Project identification

At the identification stage, when project alternatives are being considered, the overall feasibility of alternative schemes should be established and major potential hazards identified, at least in qualitative terms. An environmental profile (e.g. land use, surface and ground water, social aspects) should be outlined for each alternative.

Feasibility/appraisal

This should include hazard identification and quantitative risk analysis (QRA) for each option. The main remaining areas of uncertainty are identified. The practicability of managing major hazards is confirmed. Environmental criteria and standards are identified. The relevant local authorities should be contacted and provided with outline project information on HSE issues and conformity with legislative requirements. Environmental impact is assessed, together with the sensitivity of project economics to environmental requirements.

Project development plan

The plan should establish the design, operating and maintenance philosophy with all related hazards accounted for, and hazard control philosophy based on QRA. The plan sets out HSE guidelines and main contingency plans; reviews environmental control systems, codes of practice and procedures; reviews conformity with legal requirements; and provides further details for discussion with local authorities.

Detailed hazard analysis and operability studies are carried out. The plan identifies design and construction hazards for the development of contract strategy; identifies HSE aspects for tender evaluation; and identifies schedule points for holding design in relation to hazard control strategy. Quantitative estimates are made of all pollution possibilities, leading to the identification of options for pollution monitoring and control. A plan is developed for monitoring environmental issues during construction. The preliminary emergency management plan is prepared.

Detailed design

Key hazard control procedures are defined in relation to design and operating manuals and vendor data. Design is checked against hazard control plans. A technical review and update of QRA is carried out to confirm that all HSE philosophies are implemented. All local authority permits are obtained. The project is reviewed with community leaders.

Construction

The construction plan should be audited. Good site housekeeping is essential: planned for and strictly enforced. Review operating procedures and contingency plans. Ensure operational training is in place. Monitor environmental impact. Plan and carry out restoration of construction site. A pre-commissioning audit should be carried out to ensure that operability and safety systems are in place and effective.

Commissioning

The main operating procedures should be tested; monitor environmental impact and check effluents/emissions for conformity with specifications.

Much of the detailed work mentioned above, e.g. hazard analysis and operability studies (HAZAN/HAZOP), quantitative risk analysis (QRA), occupational health and safety studies, environmental impact assessment (EIA) and the like, is carried out by specialists. It is the responsibility of the project manager to ensure that the work is properly carried out, and that the project conforms to legislative and corporate HSE requirements at every stage of its development.

Research and thus a continuously improving scientific understanding of environmental factors can raise awkward problems for project development. Examples include a more detailed understanding of the harmful effects of substances and an improved analytical techniques for the detection of harmful substances at lower concentrations. Both may lead to the imposition of stricter regulations affecting the project. While it is difficult for companies to keep abreast of all such research, it is important that they are alert enough to developments affecting their own businesses so that action can be taken early, rather than enforced at a time when it may be extremely expensive to make changes.

Fig. 16.2 BP exploration. The Harding production platform in the North Sea (January 1996). Photograph courtesy of British Petroleum.

16.4 Decommissioning and 'abandonment'

What happens to an industrial facility when its useful life is over? In broad terms it constitutes a potential environmental hazard and, in fact, abandonment as such is only rarely possible.

An offshore platform provides an example of the regulatory framework applying to decommissioning. According to the UN Convention on the Law of the Sea (1982), it should be removed. The costs of removing what are frequently huge structures are likely to be extremely high. Possibly these costs were not fully taken into account when the structure was planned (DCF again!), so that there is considerable pressure to ameliorate the UN principle of complete removal. The International Maritime Organization (IMO) has published guidelines aimed at producing a workable compromise for the UK continental shelf, where many of the first generation of offshore structures are now nearing the end of their productive life. These guidelines indicate that an offshore structure may be permitted to remain in place subject to case-by-case evaluation of (*inter alia*):

- potential effect on safety of navigation;
- potential effect on the maritime environment, including the effect of deterioration of the structure;
- risk that the structure, or parts of it, will change position;
- cost, technical feasibility and risk of injury to personnel if the structure is removed;
- possible new uses for the structure if it remains in place;
- agreement by the relevant state authorities that the structure may remain in place.

The guidelines specify standards which must be taken into account when deciding on the complete or partial removal of an offshore structure. The standards are to be applied to existing as well as future installations, but – with effect from 1998 – no installation will be permitted unless its design and construction are such that complete removal at the end of its useful life is feasible (Fender, 1994).

The trends in regulatory thinking illustrated by the IMO guidelines show that the project leader's task in policy making is increasingly oriented towards careful consideration of long-term environmental issues. This applies to other types of project too, both onshore and offshore.

16.5 Emergency preparedness

An aspect which is frequently neglected in planning the development of a project is 'emergency preparedness'. This appears as the twelfth item in the *Charter for Sustainable Development* and it is an important element in establishing the technical integrity of the project. However, the likelihood of the project causing an emergency is also a major concern of the neighbouring community, who are among the *stakeholders* in the project. Even though they do not have the formal power to influence project decisions which is possessed by the shareholders, stakeholders can, and do, exert pressures capable of swinging project outcomes from success to failure. For this reason, emergency preparedness has to be an important area of concern for project leaders and managers alike.

EMERGENCIES

What constitutes an emergency or a disaster has been defined in various ways, e.g.

> An occurrence (including, in particular, a major emission, fire or explosion) resulting from uncontrolled developments in the course of an industrial activity, leading to a serious danger to persons, whether immediate or delayed, inside or outside the installation, or to the environment, and involving one or more dangerous substances (HMSO, 1985).

> A situation generally arising with little or no warning, and causing or threatening death, injury or serious disruption to people, premises and services which cannot be dealt with by Fire, Police and Ambulance services operating alone. The situation will require from the outset the special mobilization and coordination of a variety of other bodies and voluntary organizations (SIESO, 1986).

> An event which afflicts a community the consequences of which are beyond the immediate financial, material or emotional resources of the community (Keller, 1989).

While these definitions – and many others which may be found – reflect the interests and emphases of their authors, emergencies and disasters share the characteristics that:

- they are unforeseen;
- they invoke extraordinary measures of management and coordination.

Industry experiences a significant number – typically in the order of hundreds – of emergency incidents every year. There is some evidence that they are becoming more frequent (Keller, 1989). Having in mind the relatively small number of people who manage industry worldwide (people like you, dear reader!), it is quite likely that you will have personal experience of an emergency at some time in your career. These are man-made emergencies. Earthquakes, lightning strikes, landslips and the like are not included. Most of the emergency situations we face are accidental. That is to say, something goes seriously wrong in some of the most carefully planned activities ever undertaken by intelligent, highly trained and strongly motivated people – us!

Planning in this context is planning to (1) prevent the 'unforeseen' happening, and (2) mitigate the consequences if it does happen. 'Preventative' planning decisions are also relevant in the 'curative' sense, which is helpful in the early stages of deciding what we should do. This is rather like dealing with insect pests in your garden: if we can destroy the eggs or the larvae before they hatch, we shall save a lot of trouble later.

One particular aspect of planning for emergencies is often overlooked. If a disaster does happen, and much property is destroyed, people are killed or injured, or profitable operations are seriously interrupted, an enquiry is set up and we are at pains to learn as much as possible so as to prevent a recurrence*. We tend, however, to be much less aware of the impact on corporate cash flow of incidents reported as disasters by the media, i.e. the press, radio, television. But the influence of the media on the public – which includes investors whose decisions are largely guided by their assessment of that mysterious entity 'public confidence' – is very great. We may feel that media influence is too great (having regard to the media's predilection for drama at the expense of accuracy), but our own feelings are irrelevant. Media impact must be taken into account in the same way as the technical and human factors which we study in preparing our plans (Regester, 1989).

EMERGENCY PLANNING AND COORDINATION

As in previous chapters, the following discussion assumes a major project situation. Similar concepts apply to small projects: it is easier to scale down one's planning principles than to scale them up.

Because our project interfaces with the outside world, we have to manage the coordination of communication across these interfaces. This can be complicated. For example, in the UK, emergency planning involves coordination with the statutory emergency services (Police, Fire, Ambulance), local and national government agencies, and a number of voluntary services such as the Red Cross. In addition, there will be in-company lines of communication to be set up and managed, the local community and, of course, the media. It is easy for the number of focal points for communication to run into double figures.

Most large-scale industry operations are ultimately the responsibility of big corporations. Each operational site's emergency planning has to be coordinated with corporate headquarters' emergency planning. It is the responsibility of corporate HQ to establish emergency planning policy and strategy, and the responsibility of each site to ensure compliance by each project at the site. Some of the key issues at this level are as follows:

- What kind of events fall within the corporate definition of 'emergency'?
- What design methodologies should be implemented so as to minimize risk at the process concept and plant design phase?
- What training should be provided for operators, maintenance staff and management? How should this be practised and audited?

* The kind of legal system in force affects the likelihood of effective learning from disaster enquiries. If the main motive of the enquiry is to apportion blame, we are not likely to learn much from the enquiry. The primary motive should be to find out what caused the disaster, and only when that is known, to see who, if anyone, is to blame.

- What is corporate policy on communication, inside and outside the organization? How should this be implemented?
- What is corporate policy on minimizing business exposure and protecting the organization's assets in the event of an emergency?

Such a list can easily be extended (e.g. by a Corporate Emergency Coordination working party) to build up a checklist of strategies for development. Some will be developed in detail, others in outline, for not all eventualities can be planned for in a totally rigorous way. All the departments in the corporation likely to be affected by an emergency should be represented in such a working party, and at a sufficiently senior level that the working party's recommendations will be implemented throughout the organization.

In practice, the outcome of the working party's work will be corporate emergency planning documentation, such as a manual which will establish the framework for detailed on-site emergency planning and for coordination with off-site centres.

Planning must be in conformity with whatever statutory regulations apply to hazardous industrial operations. In the UK these are the Control of Industrial Major Accident Hazards (CIMAH) regulations, conformity with which is monitored by the Health and Safety Executive. Similar procedures are in place in all industrialized countries, and are being set up in developing countries. Differences in national legislation may result in differences in the procedural approach, but the principles of planning for emergencies are internationally similar.

EMERGENCY PLANNING 'CULTURE'

Some checklists for emergency planning (useful also for auditing the efficacy of existing plans) are available, e.g. those of the International Safety Rating System (ISRS) originating from the USA. Taking the latter as an example, the ISRS uses a detailed questionnaire to elicit the operating site management's preparedness for an emergency. These 'self-assessments' are subsequently validated by independent specialist auditors using interviews and observation to rate the site against a model representing an appropriate level of preparedness.

It is worth noting that the model does not call for *maximum* preparedness in all aspects of the site operations. In practice, this would never be feasible. The model establishes a yardstick for *optimality*. This means that, in common with other systems for rating optimal performance (rating project decisions is one), the system will be affected by corporate 'cultural' factors. Consider for example:

- **'Near miss' reporting.** A 'near miss' has the characteristics of an incident, without having the consequences. A 'near miss' is therefore more easily concealed if the incident would be considered blameworthy. But it could provide a lot of useful information, which might prevent the occurrence of a serious incident in future. So we need a 'culture' which encourages data to be made available on a blame-free basis.
- **Event concatenation.** Many emergency incidents arise from a chain or concatenation of events, each one of which is insignificant in itself. In these cases, the incident is often attributed to operator error (the operator being the person nearest the risk), when in fact the concatenation is due to policy decisions taken in the early stages of concept or design (Kletz, 1985).
- **'Homeostasis'.** According to this concept, people accept a certain level of risk in everything they do as a condition of being human. The types and degrees of risk which a person considers acceptable at any time and place are determined by culture (Adams, 1985). So training and risk reduction procedures developed and practised successfully in one

corporate culture are, like the technology to which they refer, not always directly transferable across cultures.

These examples highlight the crucial problem in managing emergency planning: even when safety and the avoidance of major loss are well understood and wholeheartedly accepted as the underlying motives, we never have unlimited resources to spend on 100 per cent security. We have, in a word, to *manage* the risk.

16.6 Summary

The potential impact of a project on the environment and the community is an increasingly important item on the agenda of project sponsors, planners and managers alike. The difficulty of relating project development to economic development as a whole leads to some opacity in the social and political aspirations which shape regulatory aims, and colours project appraisal. A consequence is that well-developed quantitative techniques, while essential in project development, do not in themselves provide a complete solution to environmental and community issues. In particular, the implications of abandonment or de-commissioning of project facilities at the end of their useful life must be considered early in project development.

Emergency preparedness is another environmental and community issue which demands early consideration, not least because of the impact of media attention (on the company, not just the project) should an emergency situation arise. Emergency preparedness involves several interfaces with the world outside the project, and highlights the importance of effective communication as well as careful planning and implementation of the project.

17
Project Reappraisal

17.1 Introduction

PURPOSE

Project reappraisal comprises a set of procedures for examining a project – or a particular phase of a project – in order to learn how to improve management of future project work. Project reappraisal also:

- helps to keep project cost estimating data up to date (e.g. equipment and material costs, cost factors, manhour rates and productivity trends);
- provides an opportunity for valuable staff training.

APPLICATIONS

Project reappraisal covers all the work we do to revise a project in view of improving its chances of success, i.e. of achieving the project objectives on time and within budget. There are four categories of project reappraisal:

- **Project reshaping** is a fundamental re-examination and revision of project objectives, which is normally carried out at the conceptual (or 'identification') stage of project development.
- **Critical review** of the project status and progress is typically carried out during feasibility studies and design, with a view to minimizing risks and enhancing the technical integrity and economic performance of the project through improved engineering.
- **Technical audit** is a formal critical review usually undertaken once, or more often, during the project's implementation phase in order to monitor conformity with the implementation plan and to assist the project manager in applying effective project controls.
- **Post-project evaluation** is a review of the total project usually undertaken some time after project completion, when normal operation has been established. The purpose is to identify particular aspects of project management decision making which should be endorsed, treated with reservation or rejected for future projects.

The basic technique for applying each category of project reappraisal is the same:

1. compare the probable or actual outcome of a decision with what was originally planned or intended;

2. where there is a mismatch between the 'actual' situation and the 'plan', identify the causes;
3. recommend action to correct current or prevent future mismatches.

BENEFITS

A project reappraisal provides an independent review of the project, with a status report on project performance. The report identifies problems, highlights key issues and their influence on the project, and makes practical recommendations for improving project performance. With the aid of project reappraisal the project sponsor (the 'client') should be able to:

- anticipate problems rather than react to them;
- confirm that the client's requirements are understood, defined and achievable;
- manage problems and risks so as to eliminate or reduce their influence on the project;
- be reassured that effective planning and control is being used;
- evaluate opportunities and options for benefiting from more or less involvement in the project;
- get an independent opinion on the performance of the project team.

AUDIT TRAILS

The basic technique of project reappraisal is adapted to suit each category of application. However, in all cases, applying the technique is facilitated when an audit trail already exists.

An audit trail is a documented record of all significant project decisions taken throughout the project. The audit trail is typically based on minutes of project meetings, project correspondence records, log books and project reports (including any previous project reappraisal reports), all of which must be suitably referenced for prompt and accurate retrieval. The trail allows project outcomes to be traced back to their origins easily and with the minimum confusion.

If you intend to use project reappraisal systematically, you should try to build the 'audit trail' concept into all your project organization procedures.

17.2 Generic reappraisal methodology

The procedures described in this section may be used as the basis for work in all reappraisal categories. Examples of adaptation for the specific categories are described in Section 17.3.

WHEN TO UNDERTAKE REAPPRAISAL

Reappraisal should be undertaken when it can yield the most useful results.

Project reshaping

Project reshaping may be undertaken at any time although it is normally carried out in the early stages of project conceptual studies (i.e. project 'identification' in the project cycle). But because the purpose of reshaping is to investigate fundamental revision of project objectives, it often leads to very far-reaching changes in the project, which may result in the loss of funds already committed if reshaping takes place after the early stages of project development.

However, project reshaping is necessary when changes in the project's (external) environment – particularly market-induced changes – are likely to affect the objectives of the project. An example is when a project has been deferred for some considerable time after completion of feasibility studies. Reshaping should be undertaken before the project is implemented.

The duration of project reshaping studies depends on the magnitude of the project under consideration – from a few days for small local projects to several weeks for large projects with international ramifications. The duration will be considerably extended in the absence of an audit trail.

Critical reviews

Critical reviews are normally carried out during design and engineering, before the design specifications and implementation plans are frozen. Critical reviews also often form part of change control procedures. If carried out on the original conceptual development of a project, a critical review corresponds to project reshaping. There must, however, be some reasonably concrete ideas to work on, otherwise it is rather like trying to reshape mud.

The purpose of critical reviews is to minimize risk, identify remaining uncertainties, assure technical integrity and improve project economics, by assessing alternative design and engineering approaches in terms of costs versus benefits, with a view to choosing the most cost-effective solution which will satisfy the requirements of the project objectives.

The duration of a critical review on any particular feature of a project is typically a few hours to a few days. Reviews are, however, normally carried out in series as the project design develops, so that the total time spent in reviews may be considerable. When the reviews are being carried out on a current project, this time must be allowed for in the project schedule. Furthermore, if the results of the reviews must be approved before work can proceed, then the time to obtain approval must also be allowed for in the project schedule.

Technical audit

Technical audit is a formal review of project status and project management methods normally made at least once during the implementation phase. The purpose of a technical audit is to provide project sponsors or bankers with an *independent* statement about the progress of the project. The audit usually reviews the project's objectives and, where appropriate, advises on the necessity for revision. Technical audits may be justified at the beginning of a large project; during procurement, during construction, in the event of a change of contract regime, or if the project schedule is deliberately shortened (i.e. by 'fast-tracking').

The basic method in technical auditing is to check selected completion dates and costs (i.e. 'milestones') achieved against those forecast in the project plan. These checks reveal the causes of potential delay or overspending so that corrective action can be taken effectively. It must be emphasized that the aim of the technical audit is to expose the project to the scrutiny of experienced *independent* engineers to determine whether the parts of the project organization are working together as planned, and to advise on the solution of potential problems. The audit team should be seen as an additional resource of experience for the project manager. The purpose of the audit is *not* to criticize the performance of the project manager. The quality of the project manager's decisions will, in any case, be shown by the overall progress of the project.

The duration of a technical audit may be from one to several weeks, depending on its scope.

Post-project evaluation

Post-project evaluation (or 'project post-mortem') is a review of the whole project usually undertaken when normal operation has been achieved and sustained. For a major project this may be up to 2 or 3 years after project completion and handover.

A post-project evaluation amounts to a final technical audit on the project, and it has the definite aim of learning from *that* project how to improve the management of the next project.

The post-project evaluation considers each of the project phases throughout the whole project cycle. It therefore begins with the examination of project concepts, feasibility and appraisal, and continues through each stage of implementation to eventual operation.

In general, the examination is carried out under the following headings: technical, economic, financial, commercial, organizational, environmental and social. Checklists are developed under each heading to detail the specific topics to be examined on the basis of identifying the relevant lessons to be learned for application in future projects.

The duration of a post-project evaluation, provided a good audit trail exists, is typically 1–2 weeks. In the absence of an audit trail, the duration will be much longer unless the evaluation is restricted to a few aspects of the project.

PRE-PLANNING

Before beginning a reappraisal study, some preliminary planning is necessary, e.g.:

- Agree with the client (i.e. the organization requesting the reappraisal) the objectives and scope of the study.
- Agree with the client the timing of the study, the way it will be organized and the method of reporting conclusions.
- Agree that the client will make available all necessary documentation and the members of the client's staff who were or are involved with the project (or the project phase) being studied. If these staff members are not available, then, as a minimum, staff who were/are familiar with the decisions made in the course of project work must be available.
- Form the reappraisal team. The team should be kept small (e.g. usually no more than six members). The majority, if not all, of the team members should not be directly associated with the project being studied. They should, however, all have relevant experience and an appropriate level of technical qualification.
- Appoint the team leader, who will act as chairman of all reappraisal study discussions and who will be responsible for the eventual study report(s). The team leader should therefore have recognized status and authority.
- The team leader should brief the team members about the objectives, scope and timing of the study, and prepare with them the approach to be adopted and the role of each team member. It is good practice to appoint a team member as secretary, i.e. to minute discussions, and to rotate this job throughout the reappraisal study.
- Ensure that adequate facilities for carrying out the study work are available.

USING CHECKLISTS

Checklists provide a convenient way to focus attention on the particular issues about which project decisions are made. However, it is never practicable to make a complete, exhaustive checklist covering all project-related issues, because any item in a checklist is capable of generating other items specific to the project under consideration.

In making the reappraisal, the checklist items are analysed using a standard logical approach. A handy mnemonic is the old rhyme:

I had six trusty serving men,
They taught me all I knew.
Their names were What and How and When
And Why and Where and Who.

Any issue at any point in decision making can be analysed with the help of these 'serving men'. For example:

1. Objective – *what* has to be achieved (i.e. by *this* decision)?
   – *why* has this (particular objective) to be achieved?
   – *what else* (i.e. alternatives) might achieve the same result?
   – *therefore*, what should be done to achieve this objective?
2. Method – *how* is this objective to be achieved?
   – *why* must it be achieved in that way?
   – *how else* might it be achieved (i.e. alternative methods)?
   – *therefore*, how should it be achieved?
3. Time – *when* has this objective to be achieved?
   – *why* must it be achieved at that time?
   – *when else* might it be achieved (i.e. alternative timings)?
   – *therefore*, when should it be achieved?
4. Place – *where* has this objective to be achieved?
   – *why* must it be achieved at that location?
   – *where else* might it be achieved (i.e. alternative locations)?
   – *therefore*, where should it be achieved?
5. Resources – *who* is responsible for ensuring that this objective is achieved?
   – *why* must that person be responsible?
   – *who else* might be made responsible (i.e. alternative resources)?
   – *therefore*, who should be responsible? Remember that the 'responsible person' cannot be held accountable for ensuring that the objective is achieved unless he has authority commensurate with the responsibility.
6. Justification – this records the reasons why the objective must be achieved, with the purpose, causes and consequences.

This procedure confirms the *intended* or *planned* outcomes of the decisions affecting each checklist item analysed. It also identifies those aspects in which the *actual* outcome does not match what was intended or planned, and exposes the causes of the mismatch.

Item-by-item checklisting is tedious if it has to be done throughout the whole project. However, this, is often unnecessary. Bearing in mind that our first step is to identify problem areas, these will become apparent from examination of project 'S'-curves (e.g. in 'earned value analysis', Section 15.4), so that detailed checklisting limited to these areas should enable key issues to be identified for further examination.

IDENTIFYING 'LEARNING POINTS'

Mismatches highlighted in comparing 'actual' against 'planned' or 'intended' outcomes indicate where decisions about one or more checklist items could be improved. For example, suppose project work – say, welding of a structure – is found to be behind schedule although expenditures are within the budget. Based on monitoring the number of welder manhours worked, a decision is taken to assign more welders to work on the structure. Subsequently, welding progress is found to remain behind schedule.

Consider:

1. Objective – *what* has to be achieved by monitoring welding progress based on welder manhours worked? *Control of welding work progress.*

– *why?* To achieve the progress intended in the project implementation plan.
– *what else?* Monitor progress by surveying actual welds completed, tested, accepted.
– *justification:* because this provides a more direct, verifiable measure of progress.

In this example, a typical route for the reappraisal enquiry would be as follows:

1. Earned value analysis shows schedule/expenditure mismatch.
2. Project report identifies 'welding' as a problem area.
3. Reappraisal team using checklist identifies monitoring decision as the specific problem.
4. Team discussion with client staff recommends improved monitoring procedure.

ORGANIZING REAPPRAISAL

The client organization is the only source of in-depth information about the project. Reappraisal cannot be carried out unless this information is made available to the reappraisal team. Consequently, it is vitally important to obtain the client organization's full cooperation in providing project information. This information is, in principle, available from (1) the client staff who were/are directly involved with the project; and (2) the client's project documentation.

Useful information can be obtained from the client's staff only if they do not feel threatened by the reappraisal. On the contrary, the client's staff should feel able and willing to cooperate in the reappraisal work. The reappraisal team must make every effort to encourage their willing cooperation.

The scope of project documentation required depends on the scope of the type of reappraisal being undertaken. The following indicates the overall scope of project information required for post-project evaluation.

Personnel

The organization of the reappraisal should be designed to encourage frank and open discussion between the client's staff and the members of the reappraisal team. This is the only way to get usable insights into actual project decisions. A fundamental requirement is that the reappraisal process is understood to be a means of learning how to enhance future project management competence, and is in no way a means of attaching blame to scapegoats for past project management failures.

Client staff are not normally brought into the team as members (an exception is the secondment of client staff for training purposes), but will be interviewed by the team in connection with the staff's project functions. These interviews must be carefully minuted by the reappraisal team secretary.

To ensure good order and effective liaison between the reappraisal team and the client, it is useful for the client to appoint one person to act as the client's counterpart to the reappraisal team leader. All formal communication should pass only through these two people.

Documentation

The following list outlines the scope of project documentation required for reappraisal:

- **Feasibility study**
  – summary and basic technical data
  – cost engineering and economics

– proposals for further development
– conclusions: recommendations for further action.

- **Development plan**
  – development objectives
  – planning basis
  – engineering considerations
  – operating considerations
  – implementation strategy
  – cost, schedule and resources estimates
  – health, safety, environmental and quality assurance requirements.
- **Design basis**
  – design approach and philosophy
  – project objectives and constraints
  – project risks
  – project location and site requirements
  – design intentions and technical specifications.
- **Project specification**
  – general project information, scope of work, site information
  – design information
  – engineering information
  – procedures and requirements for project implementation
  – cost estimates and cost data.
- **Project implementation plan**
  – project definition, including budgets and schedules
  – resources and organization
  – contracts policy, strategy and contracting plan
  – purchasing plan
  – health, safety and environmental plan
  – systems and procedures
  – commissioning and handover plan.
  – operating and maintenance requirements.
- **Project management debrief report**
  – summary and recommendations
  – development phase
  – implementation phase.
- **Monthly progress reports**
  – summary
  – design and engineering
  – project services
  – third party liaison
  – quality assurance
  – health, safety and environmental protection
  – project status and progress
  – expenditures and commitments
  – organization and project team locations.
- **WBS**
  – work breakdown structures and responsibilities for all project deliverables). *Detail depends on scope of reappraisal category.*

REPORTING

Project reappraisal reports should be kept short and to the point. Reports from critical reviews and technical audits will normally contain recommendations which require prompt decisions or approvals. These reports should make clear exactly what is being requested, who is the decision or approval authority, and at what date the decision or approval is needed. Getting a prompt response is made easier when requests are made in a standard form.

Reports from project reshaping and post-project evaluation may relate to and influence important policy issues. Policy makers are often not familiar with project technology or project management methods, so these reports must be in plain language, clear and unambiguous. Necessary technical details should be available (e.g. in annexes).

All project reappraisal reports should be considered confidential. Normally they become the property of the client organization which requested the reappraisal. If a report is to be discussed with a third party, sections on project economics and economic appraisals should be written completely separately from the rest of the report so that these sections can be removed from the copies of the report provided to the third party.

17.3 Specific reappraisal methodology

The basic approaches to carrying out the various types of reappraisal are similar. Technical audits usually go into the greatest detail, and the techniques used in auditing can be applied in other types of reappraisal, so the following discussion focuses mainly on technical audit methods.

PROJECT RESHAPING

Method

Reshaping is carried out in a series of steps:

1. **What are the real objectives of the project?** For example, is the real objective of a factory project to contribute to industrial development; to satisfy a particular market; to create employment; or something else?

   What are the essential features of the project? Could the objectives be simplified or reduced? Could the project scope be simplified or reduced?
2. **What alternatives exist which could meet the real objectives?** For example, a different policy? A different project? No project (e.g. replace manufacture by imports)?
3. **Local suitability.** Is the project appropriate for the location where it will be sited? Consider physical and climatic conditions; technology requirements; social and cultural conditions (e.g. work patterns).
4. **Finance.** Consider options for sourcing the finance required; options for structuring the financial mix; Government or private participation – what is the preferred arrangement? How will it be achieved? (e.g. joint venture, merger, etc.).
5. **Implementation.** Is the project integrated with, or does it depend on, implementation of other projects? Should the project be developed and implemented in stages? Is the proposed contract strategy appropriate (NB – balance and management of risks)? Is the proposed project organization appropriate? Should the project be smaller or larger? Would a change of scale mean a change of technology? Does the scale of the project match the availability of raw materials, utilities, human resources, market demand? (If not, what are the possibilities

of meeting any shortfall by imports, or any surplus production by exports?) Should the project be implemented sooner or later?

6. **Commercial factors.** Should the project be tied into upstream and/or downstream processing? Should it produce a partial or full range of products? Do product specifications match market requirements (NB – local vs export markets)? Could product quality be reduced without affecting product performance or sales potential? Could the project be adapted for diversification (e.g. making other products using substantially the same facilities)? Is the marketing organization appropriate?

Pitfalls

The main pitfall in project reshaping is neglecting to take account of the implications of radical revision for the project cost, time and performance criteria. Often, project reshaping is tantamount to proposing a new project without allowing sufficient time or money for proper study and appraisal of the new project.

Because reshaping is frequently initiated by government or top corporate management, project leaders and managers are likely to be pressured to go ahead with the 'new' project before the corresponding new risks have been assessed and incorporated into project management plans.

CRITICAL REVIEWS

Method

When we have at least a schematic or a flowsheet of the project, we can consider each section (i.e. each major operation) and put our six serving men to work. A convenient method is to draw on paper a matrix based on the questioning procedure set out in Section 17.2. Using this procedure helps to avoid the tendency for questioning to drift off towards irrelevancy. Recommendations can be aggregated and reported as conclusions.

The same basic critical approach is applied not only to flowsheets, but to all the project-related activities put forward in the study under review, such as marketing and organizational plans, contracting and financing proposals.

Pitfalls

When making a critical review of a study, look out for:

- assumptions not made explicit, or not brought up to date, i.e. arguments based on outdated assumptions which have never been reappraised;
- definite statements about the future, especially, for example, demand forecasts (stating 'demand for product *will* grow by *n* per cent/year'); political aspirations (notoriously unreliable as the basis for project development);
- projected rates of return (or any other measure of profitability) based on 'economies of scale' *and* export markets (or other uncontrollable market situations);
- projected sales revenue which does not take account of price/demand elasticity effects caused by the coming on-stream of the project itself;
- projected product sales which do not take account of the competition and 'customer loyalty' effects;
- overelaborate risk analysis in preliminary project evaluations;
- investment savings effected by simplifying plant design so that the project relies on the

availability of raw materials from a single source, or presupposes that the production facility need not be flexible;

- joint venture situations where the minor (e.g. technology) partner controls the greater share of marketing and sales, and there is no other significant controlling influence;
- engineering design and construction estimates (cost and schedule) when the technology is not completely developed *and* there is pressure on time;
- assumptions about 'preparedness' which affect the ability of the various parties to meet their commitments, especially:
  – site and facilities for access and development
  – trained people
  – hardware and materials including spare parts and consumables.
- assumptions about future currency exchange rates;
- shadow costing and pricing.

TECHNICAL AUDIT

Aims and scope

A technical audit has one or more of the following aims:

- to ensure compliance with regulations;
- to ensure that the project complies with plans and specifications;
- to assess the effectiveness of project management;
- to assist continuing improvement;
- to permit registration (e.g. in the special case of training audits on individuals).

The scope of an audit typically covers:

- investigation into the effectiveness of the project organization;
- verification of progress achieved against progress planned;
- evaluation of the approach to the project adopted by the client;
- examination of the influence of constraints on progress (e.g. financial, material, manpower and other constraints);
- recommendation of suitable changes when necessary.

The scope of the technical audit depends on the project contract regime. If the project is being managed under a fixed or firm price ('lump sum') regime, the audit is in principle concerned only with *time* expenditures, except in the case of changes originating either from contractors' home offices (e.g. engineering changes) or from the site (i.e. field changes). These have to be audited for both *time and money* expenditures. Similarly, in 'reimbursible' and open book contracts), both time and money expenditures should be audited.

Method

The audit consists of checking selected completion dates and costs – so-called 'milestones' – achieved against the milestones forecast in project planning. It should be obvious that, in order to be of any use for this purpose, milestones must be unambiguous and visible, whether they relate to *item* completion, *stage* completion or *project* completion.

It is often found that (1) interfaces between contractors, and their procedures for coordination, and (2) the placing of orders on vendors during the procurement phase are frequent causes of delay or overspending.

A technical audit team should be organized so that it is competent to consider four aspects of a project:

1. general business operations and management;
2. process engineering (process design and control, hazard and operability, product evaluation, environmental issues);
3. functional engineering (mechanical and electrical equipment, services, utilities, procurement and related supply issues);
4. project management (cost monitoring, scheduling, control planning, interpretation and performance, insurance, infrastructural questions relating to the site, construction and industrial relations).

'Audit' means 'he listens'. The key methodological skill in auditing is *purposeful, constructive listening.*

Main topics for study

Typically, the audit will focus on a number of broad topics where experience has shown that problems are most likely to arise, for example:

- **Management and organization** – is the project organization adequately structured and properly staffed, with reporting lines and responsibilities in place and clearly understood?
- **Project definition** – are the project aims fully established and communicated to all the people and organizations involved in the project?
- **Schedule** – are key dates realistic and identified for each stage? How is progress measured and monitored?
- **Budget** – is there a cost plan with cost control procedures?
- **Site** – are there any site-related problems, e.g. access, planning requirements etc., which affect progress?
- **Procurement** – is there a contract strategy, and was it established early enough to identify risks and allocate work to the most suitable contractors and suppliers?
- **Start-up and operations** – are the commissioning procedures agreed, and were they specified early enough? Are they taken into account in the design?
- **Communications** – are communications effective? How is cost/time/resource information transmitted within the project team?
- **Constraints** – e.g. affecting progress, such as cash flow, annual budgets, availability of suitable resources.

Pitfalls

Some important issues surface quite often during technical audits. They include:

- users (i.e. the operating function of the owner organization) not being involved in project development;
- user requirements not adequately specified in the project's conceptual stages;
- users not involved in key project approvals;
- owner's expectations unrealistic and/or subject to change;
- inadequate quality assurance, so that the final product does not satisfy the market needs;
- weak technical management (e.g. overdesign, excessive time spent on conceptual development);

- overall project objectives not well defined; not agreed with other involved parties; not understood by the project team;
- project team lacking appropriate experience and expertise;
- inadequate technical and commercial vetting of contractors and suppliers;
- inappropriate technical methods or solutions (this is relatively rare – most project failures are caused by organizational problems);
- poor communication within the project organization; responsibilities not clearly understood.
- part-time or 'amateur' project management;
- badly defined roles and responsibilities in the project organization;
- lack of direction and control in the project team (results in, for example, low productivity and missed milestones);
- contract strategy left until too late (results in risk-sharing options being neglected);
- project stages and deliverables not defined and agreed (because planning is too superficial);
- planning inadequate so that the resources required are underestimated;
- poor change control resulting in scope of work expanding and costs increasing;
- risks not well identified, so inadequate action is taken to transfer risks or make contingency allowances.

POST-PROJECT EVALUATION

A post-project evaluation is, in effect, a final technical audit. The procedure for eliciting experience (using the *six serving men* approach) is the same, and is now applied to the completed and operational project.

Method

The first major reference point for post-project evaluation is the final project appraisal, i.e. the study upon which the decision to proceed with the project was based. The evaluation scrutinizes:

- the assumptions made in the project appraisal (technical, economic, commercial, financial, etc.);
- forecasts and projections;
- assumptions on which the design was based;
- assumptions about the project's 'external environment', and the institutions involved.

These forecasts and assumptions are compared with what actually happened, establishing whether any variances were significant, why they occurred, and what can be learned from the experience. For example:

- What aspects of the project were considered in the project appraisal, e.g. technical, economic, commercial, financial, safety, environmental, social?
- What forecasts and projections were made in the appraisal; what were they based on?
- What actions or commitments (including investments) complementary to the project were considered necessary in the appraisal for the project benefits to materialize? Why were they considered necessary?
- What was the expected distribution of the project benefits?
- What institutional or organizational changes were considered necessary for successful implementation of the project? What was actually done to effect these changes?
- What functional disciplines were involved in carrying out the project appraisal?

- What Government/ministry/authority policies and procedures were in force when the project was planned, and which influenced the project? What was the actual influence?
- What working papers, calculations, drawings etc. relating to the project are available as records? One of the most important 'working papers' is the project manager's debrief report which should address, among other topics:
  - expenditure performance
  - schedule performance
  - resource performance
  - items which (with the benefit of hindsight):
    (i) were done and would be done again;
    (ii) were done and should not be done again;
    (iii) would be done differently on another project;
    (iv) could be done better (with the proposed better method).
- Were any parameters identified in the appraisal as being the key parameters on which the success of the project would be judged? What were they?
- Are any records available from before project implementation which would be helpful in assessing the local impact of the project?

Pitfalls

Probably the major pitfall in post-project evaluation is the 'post-mortem' syndrome: the hunt for a scapegoat to blame when something is found to have gone wrong. Next in line is the tendency (because post-project evaluation tends to rely on checklists to identify problem areas) to apply standard solutions to non-standard situations.

The evaluation team should keep in mind: 'What are the relevant lessons to be learned for future projects?' The project was originally formulated in the context of Government or corporate policies and procedures, and the technological and commercial framework *at that time!* The project might be different if it were formulated today, in the context of different policies, procedures, technological advances and commercial conditions. It is easy to be critical, given the advantage of hindsight!

However, the primary object of post-project evaluation is not to criticize people or apportion blame. It is, in a constructive sense, to determine the relevant lessons to be learned, so that mistakes will not be repeated, so that improved methods will be adopted, and so that future projects will be better as a result.

17.4 Summary

In project reappraisal we aim to examine past project work in order to learn how to manage better in future. There are four categories of reappraisal: *project reshaping* (typically a revision of project objectives, carried out at the conceptual stage of project development); *critical review* (typically of studies and design); *technical audit* (typically during project implementation); and *post-project evaluation* (after a period of normal operation). The basic technique is the same for all these categories.

Carrying out project reappraisal is made much easier if a documented audit trail exists. Structured questioning of previous decisions allows comparison of 'actual' with 'intended' outcomes. A mismatch locates a 'learning point' where improvement is possible. The procedure requires careful preparation and organization.

Pitfalls occur in reshaping when the implications of radical revision are neglected, and in critical reviews when we fail to make explicit the assumptions underlying studies and design work. In technical audits and post-project evaluation, most difficulties are usually due to 'people' problems, because members of the project team may feel threatened by the reappraisal. Overcoming their anxiety, by good communication and maintenance of a constructive attitude, is essential if the reappraisal is to yield useful results.

18
Project Culture

18.1 Introduction

When we speak about a culture, we usually have in mind nothing that we can specify very clearly. Culture is 'everything that humans do and do not do' (Erzenberger, 1994). In the more limited context of business, it is 'the way we do things round here' (Mole, 1990). In management, it is 'the way in which a group of people solve problems' (Hofstede, 1982).

The uniqueness of projects means that each project poses at least one new problem. In this situation people try to foresee risks (i.e. the likelihood that some event becomes a potential future problem), to make plans to cater for the risks foreseen, and to set out procedures for implementing the plans. This, we might say, lies at the heart of project culture.

Termites, whose projects are, to scale, much bigger than ours, demonstrate the use of procedures very clearly. This is how to build a termitary (*see* Fig. 18.1):

1. Make a pile of pellets (of 'carton', a mixture of wood fibre and saliva. Carton pellets stick together and eventually set solid).
2. When the pile has reached a predetermined size, look around to see if there is a bigger pile nearby. If there is, abandon your pile, and go to work on the bigger one.
3. When this pile has reached another predetermined size, explore the vicinity to see if it contains a neighbouring pile near enough to be bridged to your pile. If not, abandon your pile and search for another which does have such a neighbour, and work on that.
4. If your pile does have a neighbour close enough to be bridged, expand the top to join the two in an arch. Then carry on as before.

There are some supplementary instructions about waterproofing and orientation, but the simple procedure above produces the elaborate structure which is the termite nest (Morrison, 1979). The procedure is ideal for a project culture in which raw materials and labour resources are practically infinite, where the project sponsor and the project team are in complete agreement about project objectives, and where the labour force does its work with instinctive competence.

18.2 Checklists

In our projects it is in the early phases of the project cycle that the causes of eventual project failure often arise (Baum and Tolbert, 1985). It is also during the early phases that data for decision making are least firm, and consequently where management decisions are most likely to

1. Make a pile of pellets

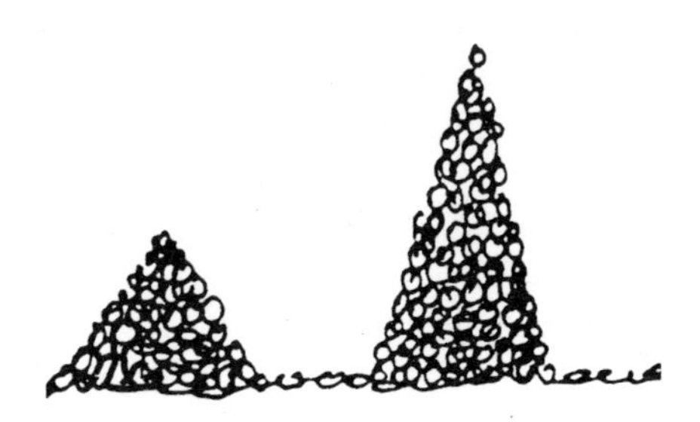

2. When the pile has reached a specified size, explore the vicinity to see if there is a bigger pile. If so, go to work on the bigger one.

3. When the pile has reached another, bigger, size, explore the vicinity to see if there is a neighbouring pile close enough to be bridged to your pile. If not, abandon your pile, search for another which does have such a neighbour, and work on that one.

4. If your pile does have a neighbour close enough to be bridged, expand the top of your pile so as to join the two in an arch.
   Then carry on as before.

Fig. 18.1 How to build a termitary: a construction algorithm.

be based on 'intuitive rationality' (Albino, 1988). In these circumstances people rely a great deal on what have been called 'pre-attentive heuristic processes', i.e. processes which determine the subject and focus of attention before we consciously apply our minds to the problem (Evans, 1989). This is facilitated by using checklists, which is why checklists appear so frequently in project management literature, and form the basis of recent project management systems such as PRINCE and RISKMAN.

In *Zen and the art of motor cycle maintenance*, Persig points out that you can organize a checklist in two ways: (1) by component parts and sublists of component parts; (2) by functions and sublists of functions. The former lists *what* should be present to make up the whole – the motor cycle, in Persig's example. The latter lists *how* it works. But neither is much use unless you are already fairly familiar with motor cycles; neither actually involves *you* (they involve only objects which are independent of you, the user); and neither concerns value judgements such as whether a checklist item is good or bad (Persig, 1988). But a checklist, in itself, does not offer anything more than completeness. Ideally, it should be:

- exhaustive, so as to be complete;
- free from overlapping items, so as to avoid double counting;
- understandable equally by the compiler and the user;
- culturally neutral.

No practical checklist achieves these ideals. In practice, our concerns are with priorities and risks, so that we can allocate appropriate and timely resources to the project. But we hardly ever have enough resources at the times when we need them. If we cannot undertake to cover all eventualities, we should attend to the most significant. And because our overriding concern is with project failure, we should prioritize our checklist with respect to the significance of the consequences of *not attending* to them. In this context, the process of decision making is not wholly rational, although it does seem to be the way in which effective decisions are made in similar circumstances (Cooper and Chapman, 1987; March and Shapira, 1987; Nutt, 1988; Chicken, 1994). This is why organizations develop *procedures*: so that people in the organization have shared understanding and interpretation of checklists.

18.3 Procedures

Talking about the rules of a particular job (it was astrology, but it could as well have been management), the character Gail Andrews in *Mostly harmless* says, 'It's just a way of thinking about a problem which lets the shape of that problem begin to emerge. The more rules, the tinier the rules, the more arbitrary they are, the better . . .' (Adams, 1992). Rules are the principles which underpin procedures. In procedures, the rules are prioritized. Both the rules and the procedures are culturally determined. Here 'culture' means management culture within business culture within the environment of 'everything that humans do and do not do'.

Procedures and project management go hand in hand. The formalization of the ways in which project work is done has become an important part of the discipline of project management. But there can be a tendency, when new (unique project) problems arise, to ignore them because there are no procedures to deal with them. In a bureaucratic organization, this will upset the view that projects are effective means of managing change (Partington, 1996).

Balance between proceduralization to avoid repeating past mistakes and proceduralization for the sake of administrative nicety is essentially a cultural balance. It is likely to tip too far if:

- the change to be effected by a project is seen in isolation, and is not linked to other areas of change in the organization;

- project management is not seen as fitting the organization's model of itself; corporate management may then resist project management;
- project management procedures involve excessive control bureaucracy (Grundy, 1993).

Corporate procedures for project management can run to several hundred pages, (perhaps confirming Ms Andrews' observation). A comparison of corporate procedures would be intolerably long-winded. We can, however, gain some insight into the ways in which people deal with their project problems by comparing project development and management under (1) centrally planned economies (CPE), and (2) market-oriented economies (MOE). The CPE environment is one in which the first of the 'management tasks' listed in Chapter 1 ('ensuring that plans and targets established by a higher authority are achieved') predominates. Tasks coming later in the list (e.g. 'making profits so as to maximize the dividend payable to shareholders' and 'optimizing operations in terms of customer loyalty to one's own organization') predominate in the MOE environment. In the following, paragraphs (a)–(g) are based on practice in the former command economies of eastern Europe (Wearne, 1994); and paragraphs (h)–(p) are based on practice in the former Soviet Union:

(a) In CPE, production or manufacturing projects are normally chosen by planners who may be quite remote from the market which determines demand for the project's output. In MOE, projects are more likely to be selected by marketing people who have to sell the product.
(b) In CPE, systematic decisions in the project selection process are, *a priori*, considered to be correct decisions. In MOE, the project selection process must take account of risk, including the risk that, during project development, market circumstances may change to such an extent that the project must be abandoned before it has been completed. (Then project expenditures will have been committed, but will never be recovered.)
(c) In CPE, project resources and finance are allocated by the central authority. In MOE, the project must compete for, contract for, and pay for all its resources and finance.
(d) Project planning in CPE is primarily concerned with following the central planning authority's instructions. In MOE, project planning is primarily concerned with exploring options and risks in order to agree objectives and priorities in authorizing the project manager.
(e) Project appraisal in CPE is based on conformity with the plan. In MOE, appraisal is based on whether the project's output will meet market demands when the project is complete.
(f) The project manager in CPE is primarily concerned to ensure that the project conforms to the central authority's plan. In MOE, he will be primarily concerned with the desirability of making changes if circumstances change. He must therefore manage the project sponsors as well as the project team. The purpose of control during the project is to influence the remaining activities so as to complete the project successfully, rather than to allocate blame for past problems.
(g) In MOE, project organizations need to show that they can learn from experience, including mistakes and failures. In CPE, we might say that project organizations must show that they do not make mistakes or have failures.
(h) It is a characteristic of management decisions that they are usually taken on the basis of information which is to some extent imperfect. That is to say, they are taken under conditions of uncertainty and risk. In the ideal CPE situation, uncertainty would be reduced. Then, risk *assessment* – essentially the quantification of uncertainty by probability – would be a priority. But in the environment of market-driven change, risk *perception* is the priority (cf. 'b' above).
(i) Organizational structures in CPE situations are steeply hierarchic, and tend to be static. Managers do not feel comfortable with the 'Western' ideal of an organizational structure

which evolves dynamically in response to (or indeed in anticipation of) perceptions of the market.

(j) In CPE organizations, communication is predominantly from the top down, although most new ideas arise at lower levels. In MOE, we like to think that we allow more 'bottom up' communication, so that new ideas have more chance of reaching and influencing decision makers.

(k) Under CPE, procedures tend to be restrictive, rigid and bureaucratic. Management is rule-based, where the rule specifies: doing *this* means not doing *that*. The alternative formulation of the rule, where you may do *this* or *that* depending on your judgement of the circumstances, is available, if at all, only to top-most management. It follows that the MOE concept of a task force, in which individual responsibilities are linked to defined limits of authority and accountability, appears in the CPE environment to be an ideal which is not, at least at present, very practical.

(l) The concept of quality assurance requires that actual performance can be assessed against a reference standard. In CPE, the reference standard is a 'given' which need not refer to the market. In MOE, the 'given' at least in principle refers to market preference. A consequence is that standards are not immutable (cf. the evolution of environmental standards).

(m) The day-to-day management of project activities is (according to MOE ideas) simplified in a project task force. Progress meetings, making reports and getting approvals, controlling interfaces and communications should be facilitated and run more efficiently, although we know that this does not always happen in practice. In the CPE tradition, project planning and design, procurement and construction are managed by separate project groups. Communication interfaces are coordinated through rigid procedures. Any change has the potential for causing serious delay.

(n) In this situation the 'blame-free' environment (which MOE protagonists say is essential if we are to learn from previous projects how to manage future projects better) is yet another ideal. Indeed, in the CPE tradition, the project organization may evolve a self-protecting structure so as to shield top managers from blame for mistakes or failure. In this case, the concept of accountability has little meaning.

(o) In the CPE environment, project policy decisions are often made with little consultation of people who have specific project expertise (cf. 'a' above). Policy makers are reluctant to authorize expenditure on studies intended to reduce uncertainty (cf. 'h' above): why spend money on 'projects' which won't go ahead?

(p) 'Stakeholder' interests under CPE are subordinated to 'shareholder' interests. The influence of the media as the voice of (some) stakeholder interests is negligible. This perhaps reinforces the customary CPE preference for reactive, curative, engineering decisions over proactive, preventative approaches to design and maintenance.

Wearne (1994) draws attention to the effect that these 'cultural' differences have in generating different approaches to the ways in which managers manage. Because there tend to be fewer levels of management in the MOE organizational hierarchy, people have to be more flexible, more self-sufficient, and more willing and able to manage risks. Managers will use their authority to ask questions, rather than to impose answers (for in CPE, authority must imply that the authority knows all the answers).

And, again in principle, management in the market economy will establish and maintain procedures only when their benefits can be shown to outweigh their costs, not because the existence of procedures is justified by a concept of public accountability. So the members of a management team should come to see themselves as the managers of *their* part of the project, accountable to all the other members of the team for performing their commitments.

18.4 Project experience: some comparisons

The only way to get at project experience is to ask practitioners, either directly or gleaned from what they publish. A few years ago I did this in an attempt to build a 'knowledge base' for a computer-driven project assistant. The attempt failed, but produced some evidence as to what experienced practitioners felt were important issues (Dingle, 1991). And here we should keep in mind that the uniqueness of projects places a premium on making good use of resources, for you never have enough to attend to everything. The important issues are therefore the issues which you cannot afford to neglect.

It turned out that there was a substantial consensus that the most important issues were:

- the acceptability of the project for development, especially regarding the response of the authorities concerned in the context of legislative requirements;
- that project objectives and priorities were clearly established;
- that project milestones were agreed and specified;
- that project engineering aspects were understood, especially in relation to equipment and materials requirements, construction, and costs and schedules coordination;
- that procedures for the control of changes were specified.

This is a rather conventional view which one might have expected from practitioners working in the same, rather stable (i.e. in the UK) business environment. There was less of a consensus about some other issues, with divergent views coming from people working in contractor organizations compared to those from owner companies. This suggests that, apart from the basics, there are potentially interesting variations in what constitutes project culture. However, it was notable that – wherever they came from – these practitioners placed technology-sensitive issues in the middle order of importance, and training-related issues were ranked low.

Project cultures are most clearly revealed in experience of projects involving technology transfer from one economic environment to another – typically from a 'developed country' transferor organization to a transferee in a newly industrializing country. Potentially instructive differences appear between the ways in which organizations and individuals perceive their problems and how to solve them.

The comparison which follows is summarized from a recent survey in which I asked experienced practitioners from several technology transferee organizations for their opinions in response to some 'culturally' oriented questions. All the practitioners come from oil/gas production and processing organizations. About three-quarters of the practitioners come from three large state-controlled enterprises (i.e. about one-quarter from each of these enterprises). The enterprises are labelled M (Malaysia), N (Nigeria), and S (former Soviet Union) below. Because the history and extent of state intervention are different for each enterprise, the respondents' perception of their overall management task may roughly be aligned with the tasks near the top of the list in Chapter 1 for group S, and near the bottom of the list for groups M and N. Moreover, the groups operate in economic climates which differ, although in each case the economic climate is in a state of change. Group M works in a boom economy where oil and gas projects contribute significantly to the boom. Group N works in an economy which has been in decline, though largely sustained by oil and gas developments. Group S works in a very large oil and gas sector which is undergoing dramatic transition from central planning to a market-oriented economy.

In the survey, the respondents agreed that project success was best measured by reference to completion within the planned budget and schedule, and in accordance with the planned duty specifications. They all agreed that, referred to these criteria, the expectations of transferee organizations were not fully realized in technology transfer projects. Overall, their expectations were only 50 per cent realized: S did rather better; N did worse than this.

The main reasons for the shortfall were said to be: (1) transferees failing to implement the information provided by the transferors; (2) transferors witholding information; (3) transferors failing to provide adequate training for transferee staff. However, S felt that transferee expectations were sometimes unrealistic.

Why do the transferee organizations fail to implement the information they get from transferors? Because – there was wide agreement on this – the transferee organizations rely on unsuitable policies for implementation, and put forward unsuitable people for training. This raises the question of what kind of training would be appropriate. There was also wide agreement that on-the-job training was best, together with specific training by technology licensors (although N preferred short 'top-up' courses to licensor training). Most respondents thought that trainees should have a Bachelor degree-level education. S preferred trainees with a Masters degree.

But no available training/education background was considered wholly adequate. Why? Because (1) there is no recognized discipline of project management relevant to technology transfer on which to base proper training programmes; (2) the transferees have no proper training policy for developing project management skills; (3) transferee company project management procedures are inadequate for technology transfer projects. N felt that the content of training programmes did not deal adequately with significant issues. S believed that cultural bias in training programmes prevented the message getting through to trainees.

Most respondents believed that projects often fail because of outside interference. The main culprits were government and government agencies. N added to these influential people associated with the community affected by the project. S blamed project financing institutions. But corporate management in the transferee organizations also caused problems by interfering, typically through changes of decision making personnel or through imposing different project objectives. This happened most often during project implementation or during preparation of the final project development plan. Almost everyone agreed that outside interference would be made less damaging if the people responsible had some training in project management.

Most of the respondents' organizations had a project management department, which supplied staff for new projects. There was considerable divergence of opinion as to the personal qualities which a project manager needed to have. On balance, training, good connections outside the organization and seniority within it seemed most desirable.

Good practice was established on the basis of comments from the parties involved with the project, and by formal post-mortems. Most organizations had written project management procedures (S less than M or N). The procedures were written by in-house staff, usually with help from outside consultants including the technology licensors. S paid the most attention to updating their procedures; N the least. But the procedures were not usually enforced – again, least so in N.

What conclusions – if any – can we draw from these observations? The respondents on the receiving end of technology transfer projects agree on how project success should be measured. They set up special project management departments, and they write project management procedures embodying good practice based on analysis of past projects, to help achieve demonstrable project success. In all this, they probably share common ground with project practitioners for whom technology transfer is less significant.

But they seem often to be careless about using their procedures. Their projects are vulnerable to outside interference which they seem unable to influence, much less control. They seem not to make best use of the information which technology transferors do provide (even though they consider that they get less than the information they believe they need) and , in consequence, their expectations are less than fully realized.

Questionable policies seem to be at the root of these problems: policies to do with the aims of technology transfer projects, with the role of project stakeholders, with the functions of project management and with the development of project management competences. In a word, with project *awareness*. Technology transfer highlights the need for project awareness, but this is not a need exclusive to projects in developing countries or transitional economies, for all projects involve the transfer of technology to some extent.

The thread which ties these issues together is 'culture' in the sense that we touched on at the beginning of this chapter (i.e. 'the way we do things round here', and 'the way in which a group of people solve problems'), and in the broader sense that culture is 'communicable intelligence'.* Communicating intelligence is what training is about, and many of the problems mentioned by our transferee respondents boil down to problems linked to training. Management culture and training are inseparable.

18.5 Training

Management culture, like other cultures, is transmitted to the rising generation through training and experience in the organization. Project management training in the tradition exemplified by central planning has two features of particular interest in the context of the observations above:

1. Management training tends to emphasize techniques rather than objectives in the design of training programmes. A mismatch between management aims and management methods sometimes results. In addition, line managers believe that training is not really their business: it is the business of training institutions or departments.
2. Senior people in training institutions (and departments) believe that the responsibility for training policy and organization should be set by some higher authority. They consider that their role should focus mainly on providing concrete, factual information, and much less on interpersonal management skills. They therefore find it extremely difficult to integrate functional with interfunctional training.

Integrated training forms a large part of, and is given increasing emphasis in, the literature of 'what the project manager needs to know' (Turner, 1996). Another re-focusing of culture is the emphasis on 'competence'. Competence goes beyond knowledge. It is the capability to *perform*, in which *knowledge, skill* (i.e. the ability to apply knowledge to a task) and *practice* (i.e. the proper application of skill in relation to some standard of 'best practice') all play a part. Competence in this sense is probably what project management practitioners have always had in mind when they think about 'what a project manager needs to know'. Practitioners rate practical 'hands-on' experience higher than academic prowess and, unsurprisingly, on-the-job training is rated as more relevant than advanced academic study. It follows that a corporate culture which favours subsidizing PhD studies for its high fliers will do more for its – and their – egos than it will for project management competence. It would do better to spend its money on providing its graduate recruits with carefully planned job experience, and encouraging its senior people to act as mentors for their juniors.

Learning and mentoring both form part of 'continuing professional development' (CPD). CPD is a necessary feature of the management culture which equips any business organization to meet the challenge of change (Burgoyne, 1992). But many organizations, and many professionals, pay more lip-service to CPD than positive action. Yet consider the following: suppose I am a senior engineer who has been practising my profession for more than 20 years. It seems to be widely

* *Encyclopaedia Britannica.*

accepted that the half-life of engineering knowhow in a mature industry is about 7 years (it will be much less in a rapidly developing industry). So, after, say, 21 years, only 12.5 per cent of what I knew when I qualified remains relevant to the practice of my profession. This equates to a 'discounted *competence* flow' at a rate of a little more than 10 per cent per year. In other words, I need to engage in CPD for something like 20 days each working year throughout my career merely to remain competent, quite apart from anything I need to do to improve my competence. (Lest accountants and others grow complacent, the same sort of argument applies to them too!) Clearly, this calls for a proper dedication to professionalism on my part, which should be supported by an appropriate management culture on the part of the organization I work for (Dingle, 1994).

Management culture finds expression in ideals (e.g. 'mission statements'), standards, norms, codes of practice etc. which are eventually embodied in procedures which give more or less formal authority to what is a corporate perception of 'best practice'. The interaction between cultural ideals, procedures and practice is illustrated by several of the observations noted above. The differences we see between practice in the 'centrally planned' and the 'Western' or market-oriented environments are coloured by differences in social as well as organizational conditions. For this reason, they are perhaps better described in terms of what practitioners find comfortable to work with, rather than in terms of right and wrong.

However, the practical issues which emerge from these observations resemble the most important problems identified for British project managers a decade ago. These were (Wearne, 1985):

- **communications** – relationships with non-technical or other technical departments that are only partly employed on a project;
- **control of costs, manhours and standards of work** – problems arising from inadequate budgeting, lack of proper information about work in hand, imposition of changes and delays;
- **contractual issues** – hasty commitments, inappropriate agreements, lack of direct control;
- **techniques** – lack of experience with systems for planning, budgeting, monitoring and reporting.

Subsequently, Wearne (1986) broke down this list into greater detail, and identified the causes of the problems. The causes (listed in Wearne's order of importance) were:

1. unpredictability of markets, governments and business conditions
2. poor planning, or plans not applied
3. reliance on promises
4. inconclusive meetings
5. departmental differences
6. cultural differences
7. no one person responsible for making decisions
8. lack of involvement and recognition
9. contractual innocence
10. engineers' lack of influence
11. lack of training
12. no confidence in improvements
13. generation gaps
14. project lessons not learned systematically
15. no-one committed to action on all the above.

Wearne (private communication, 1994) considers that his lists would not be much changed by the passage of time since his original surveys, and suggests that 'risk assessment' would be the only addition required to the original list of competences required to deal with these problems and their causes. Risk assessment is a major feature of the new competence lists for engineering project managers currently under development by the UK Engineering Council.

It is instructive to compare Wearne's lists, which identify the areas where British project engineers feel most need for training, with the short-list of 'preconditions' favouring project success mentioned in Section 1.7. The former specifies where competences should be enhanced. The latter – the preconditions – recommend what are essentially attitudes of mind which seem necessary if enhanced competences are to be successfully deployed. When, as seems to be highlighted by experience in technology transfer projects, the attitudes of mind are only loosely embodied in project policy, policy is not clearly reflected in procedures, and procedures do not sufficiently guide competence, then project management is unlikely to achieve the objectives intended by project sponsors.

18.6 Project culture in management

Recently, attention in the UK has focused on cost reduction. The 'Cost reduction initiative for the new era' (CRINE) is intended primarily to improve the economics of current and future offshore oil and gas developments (UKOOA, 1994). 'Achieving competitiveness through innovation and value engineering' (ACTIVE) is intended to cut the costs of process plant to encourage new investment (Lockie, 1996). Both initiatives are essentially the products of a project culture inspired by changing market circumstances.

The global economy is characterized by competition, and therefore by change. Capability to manage change becomes a feature of management culture, which must necessarily incorporate project culture. In corporate management, the setting of strategic objectives immediately creates a project environment: 'We aim to achieve *this* (performance/duty specification) for *this* investment (budget) in *this* period of time (schedule).'

Common failings of corporate strategy are that objectives are developed in isolation, are not communicated, or if communicated are not believed by those who have to work on them. When corporate strategy promotes a project, project culture requires us to:

- involve all the parties affected by the project;
- negotiate the project objectives and define the project;
- identify problems and risk areas before a 'go'/'no go' decision is taken;
- set risk milestones and allocate risks to the parties best able to manage them;
- create a task force;
- set clear goals for the task force, with defined responsibilities and accountabilities;
- ensure effective communication and interface management.

Project culture demands a project plan. It will have 10 key components:

- scope of work
- work breakdown and priorities
- work plan
- master schedule
- budget and cost control plan
- contracting plan
- list of the procedures to be applied, and other necessary documentation
- list of required resources: finance, manpower, matériel
- communication plan
- management plan.

For the project manager, working the project plan has five key steps:

- build and orient the team (necessary even when the team consists of only one member: you);
- clarify and communicate the project objectives, scope and WBS;
- clarify and communicate the team organization: SPR;
- clarify and communicate task dependencies;
- establish individual budgets, schedules, accountabilities.

Fig. 18.2 sums up the project management culture.

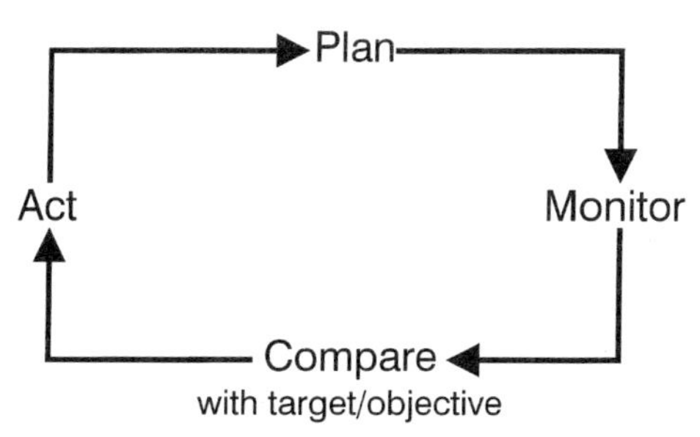

Fig. 18.2 Project control.

When we were discussing the project manager's job (Section 5.1), I said that the project manager would be likely to be rather dismissive of the motto 'Plan the work and work the plan'. I said he would prefer another motto about keeping the plan simple, but not simple-minded. Fig. 18.2 at least suggests the difference between the first motto and the second. It is that the first merely expresses an intention, which is not, in itself, enough. The second expresses an attitude – perhaps an aptitude – of mind about achieving it. The intention and the attitude come together in Fig. 18.2.

This little diagram says something to project leaders too. But the project leader is also concerned with a broader picture. He is concerned with the project in relation to the business, and in relation to the outside world and its propensity for frustrating the best laid plans, not only of the proverbial mice, but also of managers, leaders, shareholders and stakeholders. The project leader's concern is to put in place, guide and facilitate a framework within which planning, monitoring, comparing and acting can function efficiently, and to sustain it until the project is successfully integrated with the business it was intended to serve. For this, the prerequisite is project awareness. Given project awareness, project culture becomes, for project leaders and project managers alike, an instrument for (1) integrating resources for the effective management of change, and (2) learning about change in order to manage better the challenge it presents to all of us.

18.7 Summary

Because they are characterized by specifications (of objectives, budgets, schedules), projects focus the attention of those who lead them on the management of specific tasks. A fundamental task is the management of change: a project is an instrument for the management of change. It involves the management of many complementary tasks, each of which is, in effect, a sub-project. Each sub-project, in the context of the project as a whole, requires risks to be assessed and controlled; priorities to be assigned for the use of (inevitably finite) resources; appropriate expertise to be deployed; and commitment to be won and sustained.

Project leaders and managers approach these tasks and the problems which they raise from standpoints and along routes which are conditioned by their own experience and training, and the experience accumulated in their organizations of the way in which similar tasks have been (or should be) accomplished and similar problems have been (or should be) resolved. This is the 'project culture'. It is formalized in checklists and procedures.

Market competition obliges business organizations to change. Both individual and organizational experiences are shaped by the nature of the fundamental task: the kind of change the project is intended to effect. Consequently, project cultures differ. Each project invokes the opportunity to change the project culture. Consequently, project cultures evolve. They do not necessarily evolve towards a common identity, although a common ideal seems to be envisaged by project management practitioners. The ideal is embodied in *training*.

Management training involves more than communicating intelligence. It is about communicating *competence*: the capability to perform in an optimal way the management tasks which arise from the characteristic features of a project. These tasks are complex. They demand competence in planning and control techniques, methods and procedures. But the dynamics of projects in business – and in the outside world – demand in addition creativity, flexibility and only the essential minimum of bureaucracy. The mind-set which sustains this difficult balance we have called 'project culture'. Its foundation is awareness of the nature of projects and of project work: 'project awareness'. And this is what lies at the heart of the evolution of project cultures, for project leaders and managers alike.

Appendix 1: Project Implications of Fiscal Measures

A1.1 Introduction

Governments seek to maximize public revenues from industrialization projects and, at the same time, seek to pass project risk on to the developer. These aims are contradictory and result in extremely complex fiscal measures, the scale and timing of which can exert immense influence on the economic viability of a project. Furthermore, because governments sometimes act arbitrarily, fiscal measures can sometimes be illogical and unpredictable.

Thorough investigation of the potential impact of fiscal measures is always necessary before going very far with the development of any project. Specialist advice is essential. The following notes merely indicate some of the main features of taxation regimes where project leaders and managers need to satisfy themselves that fiscal distortions of project cash flows have been adequately considered.

A1.2 Downstream projects

The fiscal measures applied to projects downstream of the production of basic natural resources are in principle similar to those for other manufacturing industries. (An exception may be made for some, such as LNG projects, where sheer size and complexity demand special tax treatment.) That is to say, the project will pay corporate income tax on its profits, taking advantage of whatever allowances – such as depreciation – are permitted. Note that various different methods of calculating depreciation may lead to significant differences in the value of the allowance permitted per tax-year*. The method applicable to a particular project should be negotiated and agreed with the appropriate taxation authority, and confirmed for the life of the project.

A range of concessions may be available for projects which the Government judges will contribute to economic development. The criteria by which a project is deemed to be a 'development project' for taxation purposes are often not clearly defined, and must be confirmed by negotiation in each case. Typical concessions include:

- capital allowances, intended to encourage investment;
- tax holidays, typically limited to the early years of operation;

* For example, the 'straight-line' method assumes that the asset loses the same value during every period of its life. The 'sum-of-the-years'-digits' method assumes that the asset loses more of its value at the beginning of its life than at the end. The 'double declining balance' method assumes that the asset loses much more value at the beginning than at the end.

- tax loss carry-over;
- freedom from duty on imported equipment, subject to a variety of conditions;
- favourable treatment of infrastructural and utilities costs, again subject to conditions typically related to the project location and employment and training of local labour;
- concessions on the cost of raw materials and services;
- tariff barriers against competitive product imports;
- concessions on the repatriation of profits or on currency exchange rates (where, for example, raw materials must be purchased in a currency other than that for which product is sold).

However, tax concessions often fail to work in the way that governments intend. For example, capital allowances are usually applied to encourage investment in projects which the Government considers to be important for national economic development. But they may be used by investors simply as a tax dodge, or they may promote uneconomic project development which no rational investor would otherwise contemplate.

A1.3 Upstream projects

'Upstream' refers to the exploration, development and production of oil and gas reserves. Upstream projects are typically major contributors to national economic development (and not infrequently are indirectly the cause of national economic disturbance). The range of tax allowances, incentives and penalties applicable to downstream projects is equally applicable to upstream projects, but the latter are also subject to special fiscal treatment.

The actual fiscal regime will depend on (1) whether the Government believes it has the resources to supervise and manage the exploitation of its reserves; (2) the level of exploration, development and production. The fiscal treatment of oil and gas development illustrates this.

The fiscal treatment of oil is usually different from that of gas and associated gas projects (because investment and risk are different), but the principles are similar, i.e. the Government seeks to maximize the economic rent it can recover from exploitation, while it aims to pass all project risk to the oil/gas companies (which are supposed (1) to be willing to invest in the country, and (2) to be able to manage both country and project risk).

Thus typically, in the earliest stage of development hydrocarbon resources, the Government will give – or rather, sell – to the companies exclusive rights to explore, develop and produce in specified territories ('blocks') on payment of *royalties* and *taxes*.

Subsequently, the Government may enter into *production sharing* arrangements with the companies. In a typical production sharing agreement, the Government, acting through a state controlled enterprise, is owner and manager of the hydrocarbon reserves, with exclusive rights of exploration, development and production. Companies are invited to tender for appointment as contractor to provide all necessary risk capital and technical assistance to carry out the exploration, development and/or production work.

The proceeds from production (called 'entitlements') are classified as:

- *royalty oil/gas*, i.e. a proportion of the value of production set aside for the state;
- *cost oil/gas*, i.e. a proportion of the value of production from which the contractor is entitled to recover its costs;
- *profit oil/gas*, i.e. the remainder of the value of production which is split between the Government and the contractor.

Tax treatment of entitlements varies depending on the phase of exploration, development and production. Typically, plans and the associated budgets are submitted for Government approval annually, and reviewed quarterly. On this basis, entitlements are established quarterly.

The production sharing agreement specifies:

- the value of product to be used in calculating cost recovery;
- the contractor's liability to corporate income tax on its share of the profit oil/gas;
- the contractor's liability to 'windfall' profit tax, i.e. additional tax payable by the contractor on its sales of profit oil/gas above a predetermined base price;
- the liability to export duty on oil/gas sold by the contractor outside the Government's territory;
- the contractor's liability for contributions to training and research, typically under a linked 'technology transfer' agreement.

As governments grow more confident in their capability to manage their oil/gas reserves and the economic development which these reserves support, they may enter (through a state controlled organization) as equity partners into *joint ventures* with the oil/gas companies. The division of entitlements is then renegotiated to reflect the changed balance of investment commitments and risk. A feature of state participation in a joint venture is the option that the state's share of development costs is actually carried by the corporate partner(s) until the state's liability can be met by the state's share of production. This 'carried interest' is recoverable from the state's share of initial production. However, the state usually reserves the right to refinance its debt, and has various options for disposing of its share of production to settle its 'carried interest'.

In general, governments are anxious to get revenue from oil/gas development as early as possible. They will therefore aim to limit the amount of costs that the contractor can recover in any one year. On the other hand, they often allow accelerated depreciation in order to encourage companies to invest in development projects. These conflicting aims can be reconciled by relaxing the 'ring-fence', which usually treats each project as a separate tax entity. By relaxing the 'ring-fence', company expenditure on some projects can be offset against profits from other more fully developed projects.

This brief overview of fiscal measures is sufficient to show the many areas where choice of tax regime, assessment methods and timing can be used to manipulate project cash flows, with significant consequences for the success or failure of the project and the expectations of developers. By way of example, most of the first generation of North Sea projects overran their schedules and budgets, and failed to meet their production targets, yet were profitable thanks in part to generous tax treatment (and in part to a fortunate – and fortuitous – increase in the market price of crude oil). Currently, exploitation of marginal North Sea fields is being encouraged by favourable taxation.

Appendix 2: Investment and Growth

A business organization may consider its mission to be, for example, (1) to maximize shareholder value, or (2) to build long-term growth. The former is predicated on high returns *and* low risk. So, either project hurdle rates are set high (17–20 per cent return on net assets in the UK, compared to 9–10 per cent in Germany, second quarter 1995) or cash balances accumulate. The cash balances might then be used for acquisitions; or to attract other investors who may boost dividends to their shareholders by asset stripping. Or the cash balances attract predators. In each case, long-term growth is sacrificed. In the long-term, it seems that the net impact of takeovers is actually to reduce the return on capital of the companies making the takeover. In other words, growth by acquisition benefits companies less than growth by investment in projects.

Appendix 3: Basic Network Concepts

A3.1 Conventions

ACTIVITIES

We shall use the 'precedence network' convention, since it is more easily understood and more flexible than the 'activity-on-arrow' convention.

Each project activity is represented as a box which contains information about the activity (*see* Fig. A3.1). Relationships between the activities are shown by arrows (*see* Fig. A3.2).

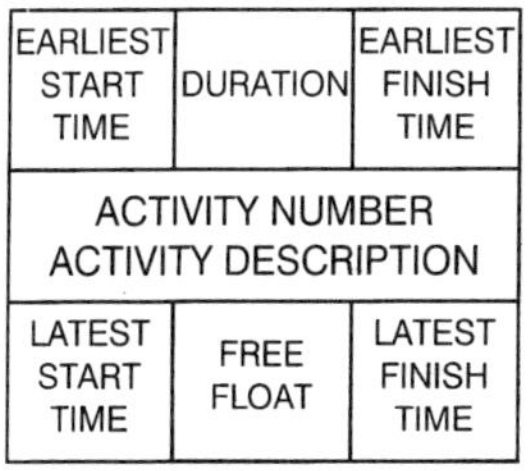

EARLIEST START TIME	DURATION	EARLIEST FINISH TIME
ACTIVITY NUMBER ACTIVITY DESCRIPTION		
LATEST START TIME	FREE FLOAT	LATEST FINISH TIME

Fig. A3.1 Standard nomenclature for precedence network activity boxes.

Each box contains the following information:

- Description and identification number of the activity.
- Duration of the activity.
- Early start time (ES), i.e. the earliest time that the activity can begin. Conventionally, ES for the first activity is 0.*
- Early finish time (EF), i.e. the earliest time it is possible to finish the activity. This will be the ES time for immediate successor activities.
- Late start time (LS), i.e. the latest time that the activity can begin without delaying the entire project. For *critical* activities LS = ES. For non-critical activities LS is later than ES and the difference is the amount of 'float' for that activity.
- Late finish time (LF), i.e. the latest time that the activity can finish without delaying the entire project. For *critical* activities LF = EF.

* Typically, work starts in the morning of the first day, and finishes in the evening of the last day of an activity's duration. If the next activity follows immediately, it will start in the morning of the following day, so a day must be added as the finish time is transferred to the start of the next activity. The ES for the first activity is taken as 1, rather than 0 as above. The effect of all this is shown in Fig. A3.3.

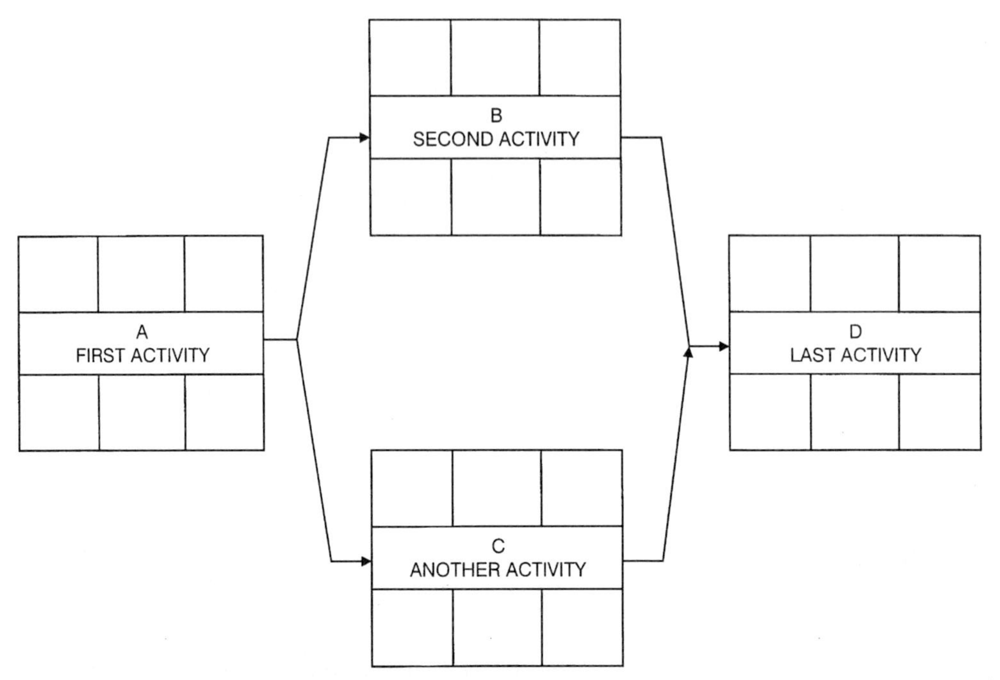

Fig. A3.2 Example illustrating relationships between activities in a precedence network.

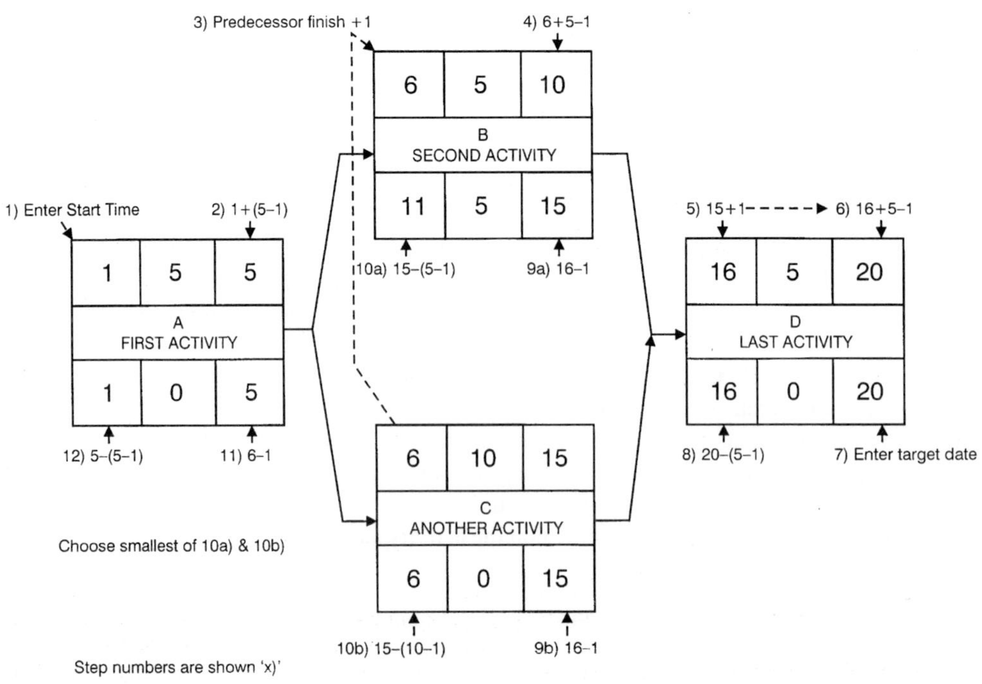

Fig. A3.3 Forward and backward pass.

DEPENDENCIES

Dependencies describe the relationships between activities:

- Finish-to-start – this is the most common. The successor activity cannot start until its predecessor has finished (Fig. A3.4).

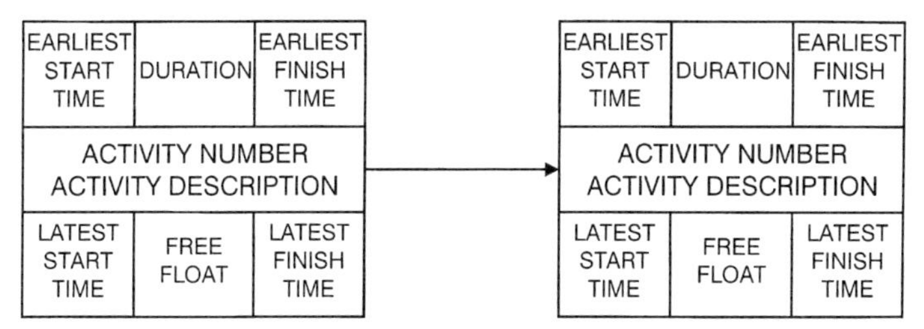

Fig. A3.4 Finish to start (FS).

- Start-to-start, finish-to-finish – both allow successive activities to overlap, as may be required in fast-tracking (Fig. A3.5). The arrow points to the dependent (or 'constrained') activity. For example, in FF the dependent activity cannot finish until its predecessor has finished.

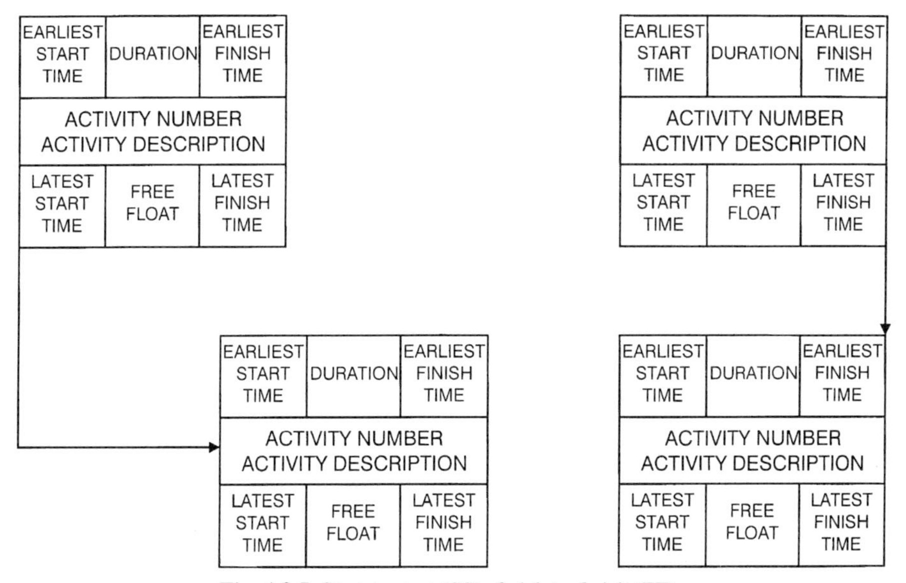

Fig. A3.5 Start to start (SS); finish to finish (FF).

- Start-to-finish – included for the sake of completion. It is difficult to think of a practical example (Fig. A3.6).

A3.2 Building a simple network

Consider a project with the activities and dependencies as shown in Table A3.1. The network looks like as shown in Fig. A3.7. Complete the network as follows:

1. Enter the ES for the first activity (no. 1).
2. Calculate the EF for the first activity. It is the ES plus the duration of the first activity.

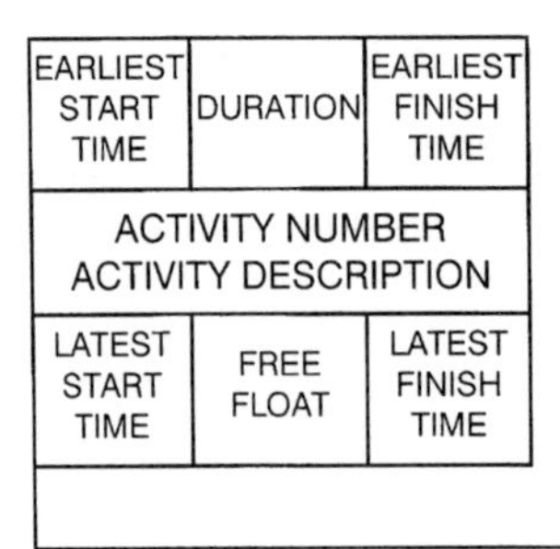

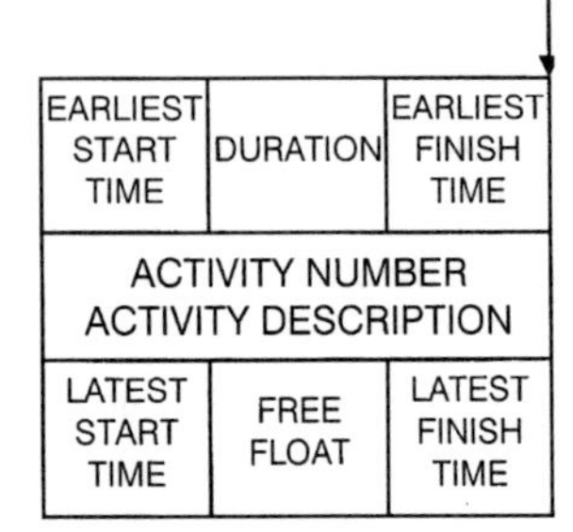

Fig. A3.6 Start to finish (SF).

Table A3.1 Calculating project duration in a precedence network

Activity	Start depends on	Activity duration
1	–	3
2	1	3
3	1	2
4	3	2
5	2 and 4	3

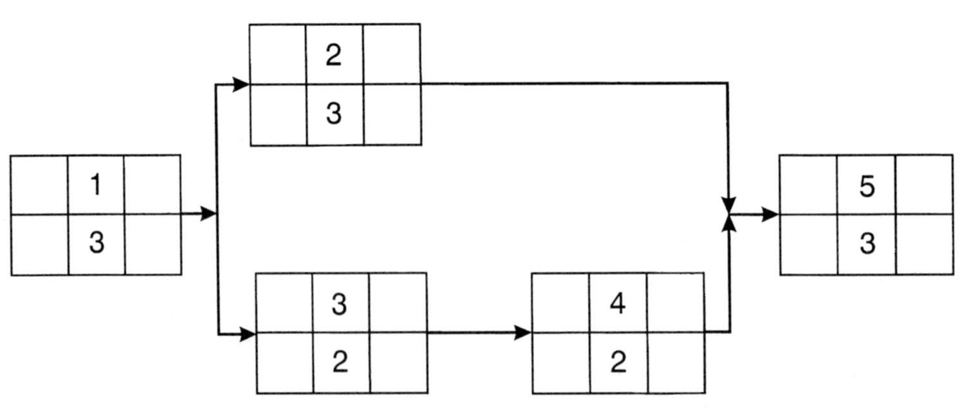

Fig. A3.7 A simplified representation of the activities. The top centre box is the activity identification number. The bottom centre box is the duration.

3. Find the next activity for which all the predecessor activities have ES and EF times. In this example, this is activity no. 2. Enter its ES, which is the largest EF of any of its immediate predecessors.
4. Calculate this activity's (i.e. no. 2) EF. It is the ES plus the duration.

5. Repeat this procedure for all the activities. The EF for the last activity is the early finish time for the entire project.
6. Find the *critical path* by working backwards through the network from the last activity to the first, following the largest EF times.
7. Enter the LF for the entire project (i.e. the target project end date). If the project is to have no float on the critical path, this will be the same as the EF for the entire project.
8. Deduct the duration of the last activity from its LF to obtain the last activity's LS.
9. Working backwards through the network, enter the LF of each activity. It will be the same as the smallest LS of any of its immediate successors.
10. Repeat this procedure for all the activities. The LS of the first activity is the latest time the project can be started and still be completed on schedule, assuming no reallocation of resources lying on the critical path.

The network should now look as shown in Fig. A3.8.

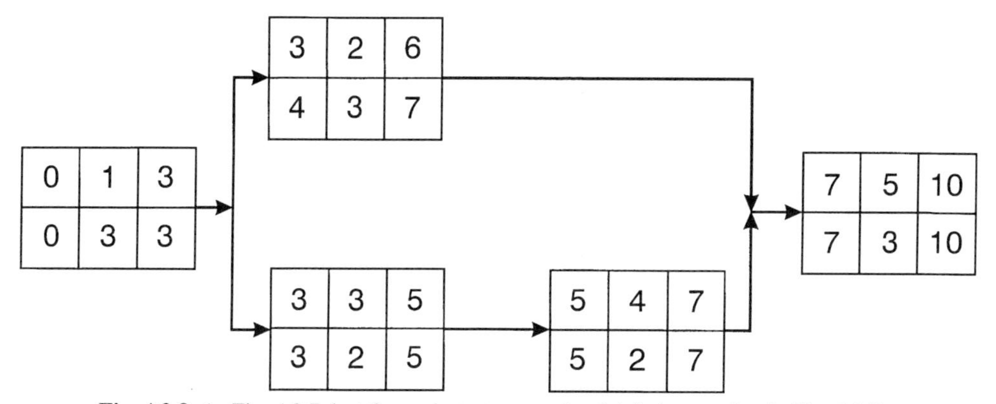

Fig. A3.8 As Fig. A3.7, but the project starts on day 0 (cf. the notation in Fig. A3.3).

A3.3 Interpreting the network

- The duration of the entire project is 10 units.
- The critical path is through activities 1, 3, 4, 5.
- Activity 2 has 1 unit of float. We could choose to delay the start of activity 2 by 1 unit without delaying the overall project.

Most project networks will be more complicated than this simple example, with more activities and more complex relationships. However, the concepts remain the same. For management reporting, the detail of complex networks is often reduced by introducing 'hammocks'. A hammock is a summary of a group of activities. The hammock's ES and EF will be the ES and EF of the first activity in the group, and the hammock's LS and LF will be the LS and LF of the last activity in the group.

Sometimes a project activity has to be related to a series of other activities. An example is the commitment of a particular resource, such as a senior engineer, to supervise a sequence of ASAP activities. The duration of any activity in the sequence will then affect the duration for which that resource is committed. Such a commitment can be specified as a hammock spanning the series of activities in question.

References and Further Reading

Chapter 1

Casson, M. 1991: *The economics of business culture*. Clarendon Press, Oxford.

Hori (a Scribe of the New Kingdom, 2nd millenium BC): Papyrus Anastasi I, BM 10247, quoted in Robins, G. and Shute, C. 1987: *The Rhind mathematical papyrus*. British Museum Publications, London.

Morris, P.W.G. and Hough, G.H. 1986: *Preconditions of success and failure in major projects*. Technical Paper No. 3, Major Projects Association Oxford.

Turner, J.R. 1993: *The handbook of project-based management*. McGraw-Hill, Maidenhead.

UNIDO (Behrens, W. and Hawranek, P.M.) 1991: *Manual for the preparation of industrial feasibility studies*. UNIDO, Vienna.

Chapter 2

Birchall, D.W. and Dingle, J. 1988: A fresh look at project preparation. *Proceedings 9th World Congress on Project Management*, INTERNET, Glasgow.

de Heredia, R. 1993: Barriers to the application of project management concepts outside entrepreneurial systems. *International Journal of Project Management* **Aug**, 131–4.

Directorate-General for Development, Commission of the European Communities, 1985: *Manual for preparing and appraising project and programme dossiers*, VIII/527/79 Rev 2.

Hutton, W. 1995: *The state we're in*. Jonathan Cape, London.

Kendrick, D.A. and Stoutjesdijk, A.J. 1979: *The planning of industrial investment programs, a methodology*. John Hopkins University Press for the World Bank, Baltimore.

Morris, P.W.G. 1994: *The management of projects*. Thomas Telford, London.

Turner, J.R. 1995: *The commercial project manager*. McGraw-Hill, Maidenhead.

The World Bank (annual) *World Bank evaluation reports*. Operations Dept, The World Bank, Washington. DC.

Chapter 3

Albino, V. 1988: Risk analysis and decision making in engineering processes. *Proceedings 9th World Congress on Project Management*, INTERNET, Glasgow.

Anand, S. and Nalebuff, B. 1985: *Issues in the appraisal of energy projects for oil-importing developing countries*. World Bank Staff Working Paper No. 738.
Channon, D.F. and Jalland, M. 1979: *Multinational strategic planning*. Macmillan Press, London.
Chapman, M. 1980: *Decision analysis*. Civil Service College Hand Book No 21. HMSO, London, 1980.
Dingle, J. 1985: Project feasibility and manageability. *International Journal of Project Management*, **3**(2), 94–103.
Harrison, P. 1983: *Operational research – quantitative decision analysis*. Mitchell Beazley, London.
Leigh, A. 1993: *Perfect decisions*. Arrow Business Books, London.
Prescott, B.D. 1980: *Effective decision-making*. Gower, Farnborough.
Rutman, L. 1976: Planning project evaluation. *UNESCO SS.76/WS/11*. Paris, France.
Simon, H.A. 1981: *The sciences of the artificial*, 2nd edn. MIT Press, Cambridge, MA, 36.
Szekeres, S. 1986: Considering uncertainty in project appraisal. *Economic Development Institute Training Materials 030/088* (Nov).
Waddington, C.H. 1977: *Tools for thought*. Granada, St Albans, 208.

Chapter 4

Chicken, J.C. 1994: *Managing risks and decisions in major projects*. Chapman & Hall, London.
Cooper, D.F. and Chapman, C.B. 1987: *Risk analysis for large projects: models, methods and cases*. John Wiley, London.
Franke, A. 1987: Risk analysis in project management. *International Journal of Project Management*, **5**(1), 29–34.
Isaac, I. 1995: Training in risk management. *International Journal of Project Management*, **13**(4), 225–9.
King, R. 1990: *Safety in the process industries*. Butterworth-Heinemann, Oxford.
McDermott, N. 1995: Heading for failure: there is an alternative to doing nothing. *Project Manager Today*, Jan, 12–15.
Oliver, R.M. and Smith, J.Q. (eds) 1990: *Influence diagrams, belief nets and decision analysis*. John Wiley, Chichester.
PRAM 1995: Project risk analysis and management. *Project*, Sept, 7–9.
Roberts, L. 1995: The public perception of risk. *RSA Journal*, Nov, 52–63 (and RM Aicken/ Lewis, correspondence in *RSA Journal* Nov 1995 and Jan/Feb 1996 issues).

Chapter 5 (and see bibliography for Chapter 15)

Briner, W., Hastings, C., Geddes, L. *et al.*, 1996: *Project leadership*. Gower, Aldershot.
Lock, D. 1996: *The essentials of project management*. Gower, Aldershot.
Obeng, E. 1996: *All change! the project leader's secret handbook*. Financial Times Pitman Publishing, London.
Stallworthy, E.A. and Kharbanda, O.P. 1986: *A guide to project implementation*. Institute of Chemical Engineers, Rugby.

Chapter 6

Belbin, R.M. 1981: *Management teams – why they succeed or fail*. Heinemann, London.
Belbin, R.M. 1993: *Team roles at work*. Butterworth-Heinemann, Oxford.

Gaisford, R.W. 1986: Project management in the North Sea. *Project Management*, **IV**(1).
Goodhart, C. 1984: *Monetary theory and practice: the UK experience*. Macmillan, London.
Lonergan, K. 1994: Programme management. *Project*, July, 12–14.
Palmer, B. 1994: Programme management in the public sector. *Project*, Sept, 14–15.
Randolph, W.A. and Posner, B.Z. 1988: What every manager needs to know about project management. *Sloan Management Review*, Summer, 65–70.
Reddin, W.J. 1987: *How to make your management style more effective*. McGraw Hill, London.

Chapter 7

Casson, J. 1983: Getting through to people – the art and craft of communicating. *Journal of the Royal Society of Arts*, April, 271–81.
Dummett, M. 1993: *Grammar and style for examination candidates and others*. Duckworth, London.
Ehrenburg, A.S.C. 1975: *Data reduction*. John Wiley, London.
Kirkman, J. 1982: *What is good style for engineering writing?* Institute of Chemical Engineers, Rugby.
Stout, R. 1955: *The golden spider*. Bantam (the quotation is from the famous (fictional) detective Nero Wolfe).
Themerson, S. 1974: *Logic, labels and flesh*. Gaberbocchus Press, London, 40.
Tufte, E.R. 1983: *The visual display of quantitative information*. Graphics Press, Cheshire, CT.

Chapter 8

Dallas, M. 1994: Achieving value – by definition. *Project*, May, 25–6.
Falconer, D. 1994: The offsite law. *The Chemical Engineer*, 26 May, 40–1.
Frater, F.G. 1992: The 5% fog. *Project*, June, 29.
Joint Working Party of The Institution of Chemical Engineers and The Association of Cost Engineers. 1988: *A guide to capital cost estimating*. Institute of Chemical Engineers, Rugby.
Kharbanda, O.P. and Stallworthy, E.A. 1985: *Effective project cost control*. Institute of Chemical Engineers, Rugby.
Liddle, C.J. and Gerrard, A.M. 1975: *The application of computers to capital cost estimation*. Institute of Chemical Engineers, London.

Chapter 9

Allen, D.H. 1991: *Economic evaluation of projects*, 3rd edn. Institution of Chemical Engineers, Rugby.
Baum, W.C. and Tolbert, S.M. 1985: *Investing in development – lessons of World Bank experience*. Oxford University Press for The World Bank.
Bergen, S.A. 1986: *Project management – an introduction to issues in industrial research & development*. Basil Blackwell, Oxford.
Caudle, P.G. 1972: New project definition. *Proceedings Conference on The Financing and Control of Large Projects*, Institute of Chemical Engineers, Rugby.
Chauvel, A., Leprince, P., Barttrel, Y. *et al.* 1981: *Manual of economic analysis of chemical processes*. Institut Français du Petrole, McGraw Hill, London (in English).
Cooper, D.F. and Chapman, C.B. 1987: *Risk analysis for large projects: models, methods and cases*, John Wiley, London.

Gruner, E., Smith, M. and Lasserre, E. 1983: *Proceedings of the Institute of Civil Engineers*, Part 1, **74**, 79–86.
Hodder, J.E. and Riggs, H.E. 1988: Pitfalls in evaluating risky projects. *Harvard Business Review*, Jan–Feb, 128–35.
Karathanassis, G. 1985: A solution to the problem of ranking projects with different economic lives. *Business Graduate*, July, 5–7.
Merrett, A.J. and Sykes, A. 1973: *The finance and analysis of capital projects*, 2nd edn. Longman, London.
Platz, J. 1986: Management of research and development projects. In Grool, M.C. *et al.* (eds), *Project Management in Progress*. Elsevier Science, Amsterdam, 247–54.
Rogers, G.F.C. 1983: *The nature of engineering*. Macmillan, London.
Royal Society Study Group on Risk Assessment 1983: Report. Royal Society, London.
Squire, L. and van der Tak, H.G. 1981: *Economic analysis of projects*. John Hopkins University Press for the World Bank, Baltimore.
Starr, C. and Whipple, C. 1980: Risks of risk decisions. *Science* **208**, 1114 *et seq*.

Chapter 10

Eales, B.A. 1994: *Financial risk management*. McGraw Hill, London.
Hancock, G. 1989: *Lords of poverty*. Macmillan, London.
Harvey, C. 1992: *Analysis of project finance in developing countries*. Gower, Aldershot.
Kendrick, D.A. and Stoutjesdijk, A.J. 1989: *The planning of industrial investment programs, a methodology*. Johns Hopkins University Press for the World Bank, Baltimore.
Pennant-Rea, R. and Emmott, B. 1987: *Pocket economist*, 2nd edn. Blackwell, Oxford.
Ruston, C. and McNair, D. 1994: Project finance in South East Asia and China. *Petroleum Economist*, **61**(2), 2–7.

Chapter 11

Beaumont, C. Dingle, J. and Reithinger, A. 1981: Technology transfer and applications, *R&D Management* **11**, 4.
Brandt Commission 1980: *North-south: a programme for survival* (see Ch. 12 on 'Transnational corporations, investment and the sharing of technology').
Brandt Commission 1983: *Common crisis, north-south: co-operation for world recovery* (especially Ch. 5 on 'The Negotiating Process').
Channon, D.F. and Jalland, M. 1979: *Multinational strategic planning*. Macmillan, London.
Fowlston, B. 1984: *Understanding commercial and industrial licensing*. Waterlow Publishers, Oxford.
Parker, V. 1991: *Licensing technology and patents*, Institution of Chemical Engineers, Rugby.
Taylor, J.H. and Craven, P.J. 1979: Experience curves for chemicals. *Process Economics International* **I**, 1.
UNIDO 1973, and Model Forms of Agreement for the Licensing of Patents and Know-how in the Petrochemical Industry (UNIDO ref. ID/WG.336/1) 1981: Guidelines for the acquisition of foreign technology in developing countries (with special reference to technology licensing agreements).
Worth Wade, 1965: *Licensing handbook*. Advance House Publishers, Ardmore, PA.

Chapter 12

Burbridge, R.N.G. (ed.) 1988: *Perspectives on project management*. Peter Peregrinus Ltd (for the Institution of Electrical Engineers), London.
Higgins, P. 1996: Contracts for consultancy services. *Project*, Jan, 14–15.
Kubr, M. (ed.) 1986: *Management consulting – a guide to the profession*, 2nd edn. International Labour Office, Geneva.
Perelman, Ya, 1973. *Figures for fun*, 2nd edn. Mir, Moscow, 144.
SIPM (Shell Internationale Petroleum Maatschappij) 1993: *Training link* (Feb). SIPM HRD Department, The Hague.
Viney, N. 1992: *Bluff your way in consultancy*. Ravette Books, Horsham.
Ward, S. and Chapman, C. 1994: Choosing contractor payment terms. *International Journal of Project Management* **12**(4), 216–21.
World Bank 1995: *Standard form of contract: consultants' services*. World Bank, Washington, DC.

Chapter 13

Abrahamson, M.W. 1978: Contractual issues – can the contract help? In *Proceedings of the conference on management of large capital projects*. Institute of Civil Engineers, London, 97–108.
Axelrod, R. 1990: *The evolution of co-operation*. Penguin Books, London (*especially Part II*).
Anon 1985: *Contract strategy – a report on client strategies when letting construction contracts*, NEDO: Engineering Construction EDC, London (*see also: Contract strategy – a pilot study and guidelines for the management of major projects in the process industries, both available from HMSO, London*).
Birkby, G. 1993: New engineering contract – not just for engineers. *Project* **Nov**, 16–17 (*see also articles in Project, Oct 1994 and March 1996*).
Casson, M. 1991: *The economics of business culture*. Clarendon Press, Oxford.
Coxon, R. 1983: How strategy can make major projects prosper. *Management Today*, April, 30–36.
Griffiths, F. 1988: Contract strategy. In Burbridge, R.N.G. (ed.), *Perspectives on project management*, Peter Peregrinus, London.
Griffiths, F. 1989: Project Contract strategy for 1992 and beyond. *International Journal of Project Management* **7**(2) 69–83.
Hegstad, S.O. 1987: *Management contracts: main features and design issues*. Technical Paper No. 65, The World Bank, Washington.
Hutton, W. 1995: *The state we're in*. Jonathan Cape, London, 250–2.
Institution of Civil Engineers 1993: *The New Engineering Contract*. Thomas Telford, London (*a second edition is due for publication in 1998*).
Knott, T. 1996: *No business as usual*. The British Petroleum Company plc, London.
Tillotson, J. 1985: *Contract law in perspective*, 2nd edn. Butterworths, London (*especially Ch. 8 on standard forms and building and engineering contracts*).

Chapter 14

Chase, R.L. (ed.) 1988: *Total quality management*. Springer-Verlag, Berlin.
Harris, R.P. 1987: The project manager and quality assurance. In *Management of new technology*. Thomas Telford Ltd, London.

ISO 8402. IOS, Geneva.
ISO 9000 1987: Quality management and quality assurance standards – guidelines for selection and use. International Organisation for Standardisation, Geneva.
ISO 9001 Quality Systems (Model for QA in design, development, production, installation and servicing). IOS, Geneva.
ISO 9002 Quality Systems (Model for QA in production and installation). IOS, Geneva.
ISO 9003 (Model for QA in final inspection and test). IOS, Geneva.
ISO 9004 QM and Quality system Elements – Guidelines. IOS, Geneva.
ISO 9004-2 (1991) Guidelines for Services. IOS, Geneva.
Janko, A.G.P. 1995/1996: *Neftegaz*. Highbury House Communications plc, London. (*Series of articles published between Jan. 1995 and Jan. 1996 – in Russian.*)
Lock, D. and Smith, D.L. (eds) 1990: *Gower handbook of quality management*. Gower, Aldershot.
Owen, F. and Maidment, D. (eds) 1993: *Quality Assurance for Process Plant*. Institute of Chemical Engineers, Rugby.

Chapter 15

Very many books on project management are available, even in non-specialist book shops. The following is a personal, and therefore idiosyncratic, selection.

PROJECT MANAGEMENT IN GENERAL

Bergen, S.A. 1986: *Project management – an introduction to issues in industrial research and development*. Basil Blackwell, Oxford.
Knoepfel, H. (ed.) 1991: Promoting and managing projects without failures. *Proceedings of the International Symposium of INTERNET, Zurich. (Collection of papers by specialists. See also other Proceedings of INTERNET Symposia on specialized issues).*
Lashbrooke, G. 1991: *A project Manager's Handbook*. Kogan Page, London.
Lock, D. 1996: *Project management*, 6th edn. Gower, Aldershot.
Lockyer, K. and Gordon, J. 1996: *Project management and project network techniques*. Pitman Publishing, London.
Maylor, H. 1996: *Project management*. Pitman Publishing, London.
Stone, R. (ed.) 1988: *Management of engineering projects*. Macmillan Education, Basingstoke.
Turner, J.R. 1993: *The handbook of project-based management*. McGraw-Hill, Maidenhead.

PROCESS INDUSTRY PROJECT MANAGEMENT

Jones, D.S.J. 1996: *Elements of chemical process engineering*. John Wiley, Chichester, Ch. 9.
Kharbanda, O.P. and Stallworthy, E.A. 1985: *Effective project cost control*. Institution of Chemical Engineers, Rugby.
Kharbanda, O.P. and Stallworthy, E.A. 1986: *A guide to project implementation*. Institution of Chemical Engineers, Rugby.
Rase, H.F. and Barrow, M.H. 1957: *Project engineering of process plants*. John Wiley, New York, reprinted 1963. (*Historical interest.*)
Whittacker, R. 1995: *Project management in the process industries*. John Wiley, Chichester.

Chapter 16

Adams, J.G.U. 1985: *Risk and freedom: the record of road safety legislation*. Transport Publishing Projects, Cardiff. (*See also Self-delusion keeps the chips going down, The Independent*, 9 July 1990.)

Beckerman, W. 1995: *Small is stupid – blowing the whistle on the greens*. Duckworth, London, 126–7.

Fender, J. 1994: Removal of offshore installations. *Energy World*, **September 1994**.

HMSO 1985: Statutory Instrument 1984 No 1902 – Health & Safety. *The Control of Industrial Major Accident Hazards Regulations*. HMSO, London.

Jacobs, M. 1991 *The green economy*. Pluto Press, London.

Keller, A.Z. 1989: *The Bradford disaster scale*. First Disaster Prevention and Limitation Conference, University of Bradford.

Kletz, T.A. 1985: *An engineer's view of human error*. Institution of Chemical Engineers, Rugby, reprinted 1987.

Little, I.M.D. and Mirrlees, J.A. 1990: Project appraisal and planning twenty years on. *Proceedings of the World Bank Annual Conference on Development Economics*. World Bank, Washington DC. (Quoted in Beckerman, 1995, above.)

Norgaard, N. 1992: *Sustainability and the economics of assuring assets for future generations*. World Bank, Washington, DC.

Moilanen, T. and Martin, C. 1996: *Financial evaluation of environmental investments*. Institution of Chemical Engineers, Rugby.

Pearce, D. 1993: *Measuring sustainable development* ('Blueprint 3' and see other essays in the Blueprint series). Earthscan, London. (Also *Economic values and the natural world*, Earthscan, London, 1993, 54–61.)

Regester, M. 1989: *Crisis management – what to do when the unthinkable happens*. Business Books, London.

SIESO, 1986: *Guide to emergency planning*. Paramount Publishing, Borehamwood, for the Society of Industrial Emergency Services Officers.

Turner, R.K. (ed.) 1993: *Sustainable environmental economics and management – principles and practice*. Belhaven Press, London.

Chapter 17

Anon 1965: *Problem analysis by logical approach – PABLA*. UKAEA, London.

UNIDO (Behrens, W. and Hawranek, P.M.) 1991: *Manual for the preparation of industrial feasibility studies*. UNIDO, Vienna.

Boughey, G.E. and Birchall, D.W. 1986: Auditing of major engineering projects. In Grooll, M.C. *et al.* (eds), *Project Management in Progress*. Elsevier Science (North Holland), Amsterdam.

Cavallone, S. 1987: Auditing large plant engineering projects. *International Journal of Project Management* **5**(1), 39–43.

Duffy, P.J. and Thomas, R.D. 1989: Project performance auditing. *International Journal of Project Management* **7**(2), 101–4.

Edgeworth Johnstone, R. 1965: *A concise guide to chemical engineering practice*. University of Nottingham, Nottingham.

Frame, Sir A. 1988: Project management: a client's view. In Burbridge, R.N.G. (ed.), *Perspectives on project management*. Peter Peregrinus, London, 7.

Hayfield, F. 1986: Project successes and failures. In Grool, C. *et al.* (eds), *Project management in progress*. Elsevier Science (North Holland), Amsterdam.

Isaac, I. 1995: Training in risk management. *International Journal of Project Management* **13**(4), 225–9.

Kharbanda, O.P. and Stallworthy, E.A. 1983: *How to learn from project disasters*. Gower, Aldershot.

McDermott, N. 1995: Heading for failure: there is an alternative to doing nothing, *Project Management Today*, **Jan**, 12–15.

Pruitt, G.C. and Park, C.S. 1991: Monitoring project performance with post-audit information, *Engineering Economist* **36**(4), 307–35.

Shannon, D. 1994: Know thine adversary – a trip through the project auditing maze. *Project*, **Nov**, 8–10.

The World Bank 1979: Course Note CN862, EDI Department. The World Bank, Washington, DC.

Chapter 18

Adams, D. 1992: *Mostly harmless*. Heinemann, London.

Albino, V. 1988: Risk analysis and decision-making in engineering processes – a conceptual review of the state of the art. *Proceedings 9th World Congress on Project Management*, INTERNET, Glasgow.

APM 1994: *Association of Project Managers Yearbook*. APM High Wycombe.

Baum, W.C. and Tolbert, S.M. 1985: *Investing in development – lessons of World Bank experience*. Oxford University Press for the World Bank.

Burgoyne, J. 1992: Creating a learning organisation. *RSA Journal*, Apr, 321–32.

Carter, B., Hancock, T., Morin, J.-M. *et al.*, 1994: *Introducing RISKMAN methodology*. NCC Blackwell, Oxford.

Chicken, J.C. 1994: *Managing risks and decisions in major projects*. Chapman & Hall, London.

Cooper, D. and Chapman, C. 1987: *Risk analysis for large projects: models, methods and cases*. John Wiley, London, Ch. 6.

Dingle, J. 1991: Cultural issues in the planning and development of major projects. *International Journal of Project Management*, **9**(1), 29–33.

Dingle, J. 1994: Training Managers to Manage Technology Transfer. *Technology Transfer & Acquisition Seminar*. Malaysian Technology Development Corporation & Petronas, Bangi, Selangor, 12–13 Dec.

Eiser, J.R. and v.d. Pligt, J. 1988: *Attitudes and decisions*. Routledge, London.

Erzenberger, H.M. 1994: The great migration. In *Civil War*. Granta Books, London, 134–5.

Evans, J.B.T. 1989: *Bias in human reasoning – causes and consequences*. Lawrence Erlbaum Associates, Hove.

Grundy, T. 1993: *Implementing strategic change: a practical guide for business*. Kogan-Page, London. (Quoted in Partington *op. cit. supra.*)

Handy, C. 1985: *Gods of management*. Pan, London.

Hofstede, G. 1984: *Culture's consequences – international differences in work-related values*. Sage Publications, London.

Lockie, M. 1996: Getting active. *The Chemical Engineer*, 25 July, 27–9.

Mallagh, C. 1988: The inherent unreliability of reliability data. *Quality & Reliability Engineering International*, **4**, 35–9.

March, J.G. and Shapira, Z. 1987: Managerial perspectives on risk & risk taking, *Management Science* **III**, 11.

Mole, J. 1990: *Mind your manners*. The Industrial Society, London, 160.
Morrison, P. 1979: *The Listener*, 23 August.
Nutt, P.C. 1988: The effects of culture on decision making. OMEGA. *International Journal of Management Science* **16**(6): **553–67.**
Partington, D. 1996: The project management of organisational change. *International Journal of Project Management* **14**(1), 13–21.
Persig, R.M. 1988: *Zen and the art of motor cycle maintenance*. Vintage, London.
PMI 1987: *Project management body of knowledge*. Project Management Institute (revised 1994).
Randolph, W.A. and Posner, B.Z. 1988: What every manager needs to know about project management. *Sloan management review*, Summer, 65–70.
Schlick, J.D. 1988: Developing project management skills. *Training & Development Journal*, May, 20–8.
Seiler, R.K. 1990 : Reasoning about uncertainty in certain expert systems: implication for project management applications. *International Journal of Project Management* **8**(1), 57–9.
Thornberry, N.E. 1987: Key areas of skills-based training for project managers. *Training & Development Journal* (Oct) 60–2.
Trompenaars, F. 1993: *Riding the waves of culture*. Economist Books, London.
Turner, J.R. 1996: Editorial in *International Journal of Project Management* **14**(1), 1–6.
UKOOA, 1994: *CRINE – Cost Reduction Initiative for the New Era*. Report published on behalf of the United Kingdom Offshore Operators Association by the Institute of Petroleum, London.
UNIDO (Behrens, W. and Hawranek, P.M.) 1991: Manual for the preparation of industrial feasibility studies. UNIDO, Vienna.
Wearne, S.H. 1994: Preparing for privatised project management. *International Journal of Project Management* **12**(2), 118–20.
Wearne, S.H. 1985: *International Journal of Project Management* **3**(3), 150–52.
Wearne, S.H. 1986: Training needs of individuals. *IChemE Conference on Project Management Training Today & Tomorrow*, London (19 June).
World Bank Annual Reviews of Project Performance Audit Results, 1985 *et seq*.

Appendix 2

Coyle, D. 1994: Article in *The Independent*, 11 October.
Davis, J. 1995: Article in *The Independent*, 30 March.
Euromoney 1995. March, 110–11.
The Guardian 1996: Reports of annual conference of the Royal Economic Society, 4 April.
Hutton, W. 1995: *The state we're in*. Jonathan Cape, London.
Hutton, W. 1996: Journal of the Royal Society of Arts, March, 35.
Searjent, G. 1995: Article in *The Times*, 18 April.

Index